国外优秀物理著作
原版系列

量子物理的非常规方法

Uncommon Paths in Quantum Physics

[俄罗斯] Kazakov, K. V.（卡扎科夫） 著

哈尔滨工业大学出版社
HARBIN INSTITUTE OF TECHNOLOGY PRESS

黑版贸审字 08-2015-063 号

Uncommon Paths in Quantum Physics
Konstantin V. Kazakov
ISBN:978-0-12-801588-9
Copyright © 2014 Elsevier Inc. All rights reserved.

Authorized English language reprint edition published by Elsevier (Singapore) Pte Ltd. and Harbin Institute of Technology Press

Copyright © 2016 by Elsevier (Singapore) Pte Ltd. All rights reserved.
Elsevier (Singapore) Pte Ltd.
3 Killiney Road, #08-01 Winsland House I, Singapore 239519
Tel:(65)6349-0200   Fax:(65) 6733-1817
First Published 2016
2016 年初版

Printed in China by Harbin Institute of Technology Press under special arrangement with Elsevier (Singapore) Pte Ltd. This edition is authorized for sale in China only, excluding Hong Kong SAR, Macao SAR and Taiwan. Unauthorized export of this edition is a violation of the Copyright Act. Violation of this Law is subject to Civil and Criminal Penalties.

本书英文影印版由 Elsevier(Singapore) Pte Ltd. 授权哈尔滨工业大学出版社在中国大陆境内独家发行。本版仅限在中国境内(不包括香港、澳门以及台湾)出版及标价销售。未经许可之出口,视为违反著作权法,将受民事及刑事法律之制裁。
本书封底贴有 Elsevier 防伪标签,无标签者不得销售。

图书在版编目(CIP)数据

量子物理的非常规方法=Uncommon Paths in Quantum Physics:英文/(俄罗斯)卡扎科夫(Kazakov, K. V.)著. —哈尔滨:哈尔滨工业大学出版社,2016.1
ISBN 978-7-5603-5765-2

Ⅰ.①量… Ⅱ.①卡… Ⅲ.①量子论-英文 Ⅳ.①O413

中国版本图书馆 CIP 数据核字(2015)第 293705 号

策划编辑	刘培杰
责任编辑	张永芹　刘立娟
封面设计	孙茵艾
出版发行	哈尔滨工业大学出版社
社　　址	哈尔滨市南岗区复华四道街 10 号　邮编 150006
传　　真	0451-86414749
网　　址	http://hitpress.hit.edu.cn
印　　刷	哈尔滨市工大节能印刷厂
开　　本	787mm×1092mm　1/16　印张 13.75　字数 222 千字
版　　次	2016 年 1 月第 1 版　2016 年 1 月第 1 次印刷
书　　号	ISBN 978-7-5603-5765-2
定　　价	118.00 元

(如因印装质量问题影响阅读,我社负责调换)

Contents

Preface   iii

1   Ideas and principles   1
    Quantum world   1
       Probability waves   3
       Physical operators   6
    Noncommutative physics   10
       Moment of momentum   14
       Perturbation theory   19
    Factorization   21
       Oscillator   23
    Quantum numbers   27

2   Physics of the electron   31
    Hydrogen atom   31
       Bohr's formula   33
       Matrix elements   37
    Dirac's equation   38
       Relativistic invariance   43
    Spin one-half   46
       Pauli's theory   52
    Elementary consequences   57
       Useful definitions   61
       Positrons   63
    Fine structure   64
       Solution according to factorization   68
    Magnetic interaction   71
       Landau levels   74
       Anharmonicity   76

3   Theory of anharmonicity   79
    Model Hamiltonian   79
    Perturbation method   82
       Inclusion of degenerate levels   86
    Polynomial formalism   89
       Ensemble of anharmonic oscillators   94
       General equations   97

	Physical interpretation	**99**
	Rules for polynomials	**101**
	Quantum functions	**104**
	Other anharmonic models	**106**
	Morse potential	**107**
	Generalized Morse problem	**110**
4	**Quantum fields**	**115**
	Creation and destruction operators	**115**
	Free scalar field	**120**
	Quantization of electromagnetic field	**126**
	Fermi's ideology	**128**
	Electron–positron Dirac field	**133**
	Interaction picture	**138**
	Solution according to perturbation theory	**144**
	Normal product	**147**
	Ultraviolet divergences	**152**
	Regularization of interaction energy	**152**
5	**Radiative corrections**	**159**
	Renormalization of mass	**159**
	Anomalous magnetic moment of the electron	**165**
	On the history of radiative corrections	**172**
	Bethe's formula	**173**
	Electromagnetic shift of atomic levels	**177**
	Vacuum polarization	**185**
	Renormalization of charge	**192**
	Dirac's ideas and quantum field theory	**193**
Bibliography		**197**

Preface

*Quantum insights serve to unwind
conundrums of nature through power of mind...*

Quantum mechanics is one of the most fascinating, and at the same time most controversial, branches of contemporary science. Disputes have accompanied this science since its birth and have not ceased to this day. What is the sense of a probability interpretation of a physical phenomenon? Which approach to a quantum field theory is more consistent? How must we comprehend a quantum world? This book, leaving aside the search for spiritual content and answers to these questions, allows one to deeply contemplate some ideas and methods that are seldom met in the contemporary literature. Instead of widespread recipes of mathematical physics based on the solutions of integro-differential equations, we prefer logical and partly intuitional derivations of noncommutative algebra. The reader, having become armed with the necessary knowledge and skills from classical physics and symbolic mathematics, can thus directly penetrate the abstract world of quantum mechanics.

For exactly solvable models, we develop the method of factorization. This method, leaning primarily on Green's formalism, is applied for consideration of simple problems in the theory of vibrations and the relativistic theory of an electron. For more complicated problems, mainly related to the physics of various effects of anharmonicity, we develop the method of polynomials of quantum numbers, which enables one to systematize the calculations according to the perturbation theory. Regarding the quantum field theory and the calculation of observable radiative corrections, we rely entirely on Dirac's ideas, not on — at present — the pervasive rules of operation with a scattering matrix. Dirac's theory, possessing a proper elegance, is built on the equations of motion and is suitable for a first acquaintance with the principal problems of quantum electrodynamics, a matter of belief that remains open.

The author respectfully expresses his gratitude to John Ogilvie, who read the manuscript and made valuable comments. This book addresses a wide readership with serious enthusiasm about theoretical physics.

Konstantin V. Kazakov
Irkutsk, Russian Federation
December 2013

Ideas and principles

Quantum world

Classical physics and theory of relativity developing simultaneously with an understanding of the world of atoms and molecules, elementary particles, and fundamental interactions marked the beginning of a new era — the era of the quantum world.

Why are some substances dense and strong while others are loose and brittle? What is the cause of the spontaneous fission of uranium nuclei? Why is copper a conductor and why is glass an insulator? What is the nature of magnetite and magnetism? How does the sun work and why can solar radiation energy be converted into electricity? What are methane clathrates? What is the essence of the problem of global warming? Why are the spectra of various substances distinguishable and, moreover, discrete? For these and many other questions, classical physics fails to yield quantitative answers. To reply to the queries about the world of atoms and molecules is the mission of quantum mechanics.

Quantum mechanics is a complicated science. Many and various substances surround us; for us they are complicated and unique, but at the same time there is not great difficulty classifying them in terms of chemical elements. Water is H_2O, methane is CH_4, carbon dioxide is CO_2, and so on. Each atom represents an element of Mendeleev's table, and the table is the fruit of quantum mechanics. Elementary particles, in a sense forming the tangled "invisible" world of reactions and mutual transformations, are hiding inside the atoms. For instance, in a collision of two protons one can see the birth of the meson of π-type or K-mesons may be created, but protons disappear at the same time. Moreover, there exists a deep relation between electromagnetic radiation and particles, so the reaction of a collision of an electron and a positron can lead to the creation of γ-quanta and annihilation of the electron−positron couple. Naturally, the reverse process is also possible when a photon having sufficient energy initiates the appearance of particles. A series of these and many other phenomena from the microworld is described in the specific language of quantum mechanics.

Paying tribute to the history of the development of quantum theory, let us briefly consider a few problems in which the argument of classical physics experiences the greatest difficulties.

We begin with the problem of black body radiation. Let there be a cavity, and its hole represents a perfect black surface. A light, penetrating inside, is repeatedly scattered and absorbed. If thermal energy, which is emitted by the walls of the cavity inside, equals the energy that is absorbed by these same walls, then thermal equilibrium is maintained. How does the energy density u in the cavity depend on wavelength λ and temperature T? Planck replied to this question. Recall, according to Wien's displacement law,

$$\frac{\lambda_{max}}{c} \cdot k_B T = C = \text{const},$$

in which k_B is the Boltzmann constant, c is the speed of light, λ_{max} corresponds to the maximum of function u, and constant C has a specific dimension,

time × energy.

Planck established the general law for the energy per unit volume and per unit of wavelength interval, and he calculated constant C; as a result,

$$u(\lambda, T) = \frac{8\pi hc}{\lambda^5} \cdot \frac{1}{e^{hc/\lambda k_B T} - 1}$$

and $C = 0.2h$, and

$$h = 6.63 \times 10^{-27} \text{ erg} \cdot \text{s}$$

is the fundamental Planck constant. With the birth of h, quantum physics was born.

Another controversial problem is the photoelectric effect. Quite surprisingly, under the action of light, according to experiments by Millikan, one might eject the electrons from the metal surface. The emission of electrons is understandable, but the fact that the electron energy depends linearly on frequency ν of incident radiation is much more complicated to understand, especially from the classical point of view. Einstein resolved this conundrum and offered an elegant explanation of the photoelectric effect,

$$E_e = h\nu - W,$$

in which E_e is the electron energy and W is the work function. The intrigue of the wave−particle duality of light was thus revived.

Finally, we discuss a question related to the stability of the lifetime of atoms. According to Rutherford's experiments, most α-particles pass practically without hindrance through a foil of gold, deflected by nuclei at definite angles, which one might easily calculate by applying classical considerations. These experiments prove, firstly, that an atom is a nucleus plus electrons and, secondly, that Coulomb's law is valid at atomic distances. From Rutherford's experiments, it is impossible to evaluate the size of atom. Moreover, the planetary model is inconsistent with classical electrodynamics; otherwise, an electron moving with acceleration around the nucleus would be forced to radiate at each revolution, losing its energy, and eventually would fall into the nucleus after only one-hundredth of a nanosecond. A lifetime would not be long.

The explanation of the "paradox" of stability of atoms came from spectroscopy. Analyzing the atomic spectra, Ritz stated that the distribution of frequencies ν occurs in accordance with the combination principle

$$\nu_{sn} = \nu_s - \nu_n = \mathrm{Ry}(n^{-2} - s^{-2}),$$

in which Ry is the Rydberg constant and s and n are integers other than zero. According to the classical description, the spectral lines must be equidistant from each other. In fact, the lines converge with increasing s, which is confirmed, for instance, by the Balmer series for hydrogen atoms. Bohr, understanding that $h\nu$ represents the energy itself, proposed a quantum model of an atom, according to which the atom, being in a stationary state, does not emit or absorb radiation, and transitions possible from one state to another occur in compliance with the law of conservation of energy according to a rule

$$h\nu_{sn} = E_s - E_n.$$

If an atom, for example, emits a photon, then an electron jumps from some stationary state with energy E_s to the state that is characterized by energy E_n; in this case, the frequency of the photon equals $\nu_{sn} = (E_s - E_n)/h$. The case of absorption is interpreted analogously.

Probability waves

To a particle of the microworld, de Broglie ascribed wavefunction φ. Realizing this determination in a literal sense, de Broglie introduced the relations

$$E = \hbar\omega \text{ and } \mathbf{p} = \hbar\mathbf{k},$$

which connect energy E and momentum $\mathbf{p}$ of a freely moving particle with frequency ω and wave vector $\mathbf{k}$ of a plane wave

$$\varphi \sim e^{i(\omega t - \mathbf{k} \cdot \mathbf{q})},$$

in which $\omega = 2\pi\nu = c|\mathbf{k}|$ and $\hbar = h/2\pi$. In other words, apart from the material, there is "something" that at each point $\mathbf{q}$ of space provides some information about the particle at the moment of time t. Born proposed the probability interpretation for function φ. These waves are probability waves, for instance, the waves of probability where the particle is located at a concrete point of space or where the particle has a concrete value of energy. Despite his phenomenological explanation of the photoelectric effect, Einstein contested vigorously against the probability interpretation of a wavefunction: "God does not play dice." Despite all its successes, Einstein, as is well known, treated quantum mechanics cautiously.

Depending on a *representation*, a wavefunction can be specified as the function of corresponding variables; for instance,

$$\varphi(\mathbf{q}), \chi(\mathbf{p}), \text{ and } \theta(E)$$

are images of a wavefunction in coordinate, momentum, and energy representations, respectively. To not indicate explicitly the chosen representation, it is convenient to express the states of quantum-mechanical systems through abstract vectors in a *separable* Hilbert space. According to Dirac's notation, to a vector that characterizes some state n we ascribe symbol $|n\rangle$; furthermore,

$$\langle n|$$

is the vector that is the complex conjugate of $|n\rangle$,

$$c \cdot |n\rangle$$

is the multiplication of vector by a complex number c,

$$\langle n|m\rangle$$

is the scalar product of two vectors $|n\rangle$ and $|m\rangle$, and

$$\langle m|G|n\rangle$$

is the matrix element of physical quantity G between states n and m. Vectors $|n\rangle$ and $\langle n|$ are called ket and bra, respectively.

Any physical vector $|\Psi\rangle$ is expressible as a series expansion in terms of *orthonormal* states $|1\rangle, |2\rangle, \ldots, |\ell\rangle, \ldots$, which in a set form a complete basis, and in a linear manner

$$|\Psi\rangle = c_1|1\rangle + c_2|2\rangle + \cdots + c_\ell|\ell\rangle + \cdots,$$

in which $c_1, c_2, \ldots, c_\ell, \ldots$ are the pertinent coefficients; basis vectors $|\ell\rangle$ fail to be related to each other through any relation of linear type — they are linearly independent. The orthonormality condition means that

$$\langle k|\ell\rangle = \delta_{k\ell}, \quad \ell \text{ and } k = 1, 2, \ldots,$$

that is, the vectors are mutually orthogonal and each of them is normalized to unity. From a physical point of view, the state of a quantum-mechanical system is a *superposition* of all possible for this system state, and each of which gives its own contribution with a weight that is defined by corresponding coefficient c_ℓ. Amplitude c_ℓ thus has a purely probability character, and quantity

$$|c_\ell|^2$$

can be interpreted as the probability of state ℓ. Assuming that ket $|\Psi\rangle$ is normalized to unity, one might readily obtain this equality,

$$\sum_\ell |c_\ell|^2 = 1,$$

which proves that the total probability, as one should expect, equals unity; the event that the system occupies one available state, obviously, represents a certain event.

In common use, the coordinate representation is applied to the problems of quantum mechanics. In this case, the scalar product of wavefunctions is given by this expression

$$\int_M \varphi_n^*(\mathbf{q})\varphi_m(\mathbf{q})\mathrm{d}\mathbf{q},$$

in which M is the manifold, in which functions φ_n and φ_m, corresponding to states n and m, respectively, are defined and $\mathrm{d}\mathbf{q}$ is an element of volume. Quantity

$$|\varphi_n(\mathbf{q})|^2\mathrm{d}\mathbf{q},$$

according to Born's probability interpretation, represents the probability that the values of coordinates of a system lie in an interval between $\mathbf{q}$ and $\mathbf{q} + \mathrm{d}\mathbf{q}$. Thus, diagonal matrix element $\langle n|G|n\rangle$, which is equal to

$$\int_M \varphi_n^*(\mathbf{q})(G\varphi_n(\mathbf{q}))\mathrm{d}\mathbf{q},$$

should be understood as the average value of physical quantity G in state n; if G is simply the coordinate function, then

$$\langle n|G|n\rangle = \int_M G(\mathbf{q})|\varphi_n(\mathbf{q})|^2\mathrm{d}\mathbf{q}.$$

The general formula for an arbitrary matrix element in the coordinate representation, by definition, has the form

$$\langle n|G|m\rangle = \int_M \varphi_n^*(\mathbf{q})(G\varphi_m(\mathbf{q}))\mathrm{d}\mathbf{q}.$$

Analogous considerations are applicable for the case of the momentum representation. For instance,

$$|\chi_n(\mathbf{p})|^2\mathrm{d}\mathbf{p}$$

is the probability to detect the momentum of a system in the vicinity of point $\mathbf{p}$ in the element of volume $\mathrm{d}\mathbf{p}$ of the momentum space, and

$$\int_{M'} \chi_n^*(\mathbf{p})(G\chi_n(\mathbf{p}))\mathrm{d}\mathbf{p}$$

represents the average value of quantity G in the state with wavefunction χ_n that is defined in manifold M'.

Physical operators

Suppose that q and p are the canonically conjugate coordinate and momentum, $\varphi(q)$ is the wavefunction in the coordinate representation, and $\chi(p)$ is the same function, but in the p-representation; quantities $\varphi(q)$ and $\chi(p)$ are related between each other through the Fourier integral

$$\chi(p) = \frac{1}{\sqrt{2\pi\hbar}} \int_{-\infty}^{+\infty} \varphi(q) e^{-i(pq/\hbar)} dq.$$

The average value of momentum p, on the one side, is expressible as

$$\langle \chi|p|\chi \rangle = \int_{-\infty}^{+\infty} \chi^*(p) p \chi(p) dp,$$

and on the other side,

$$\langle \varphi|p|\varphi \rangle = \int_{-\infty}^{+\infty} \varphi^*(q) (?) \varphi(q) dq,$$

through which symbol "?" designates the operator of momentum in the coordinate representation. Our task,[1] taking into account that

$$\langle \chi|p|\chi \rangle = \langle \varphi|p|\varphi \rangle$$

is to determine the explicit form of "?."

So, using the expansion of $\chi(p)$ in the Fourier integral, we have

$$\langle \chi|p|\chi \rangle = \int_{-\infty}^{+\infty} \frac{dp}{2\pi\hbar} \int_{-\infty}^{+\infty} \varphi^*(q') e^{i(pq'/\hbar)} dq' \int_{-\infty}^{+\infty} \varphi(q) p e^{-i(pq/\hbar)} dq.$$

Putting

$$p e^{-i(pq/\hbar)} = i\hbar \frac{\partial}{\partial q} e^{-i(pq/\hbar)}$$

and taking into account the boundary conditions $\varphi(\pm\infty) = 0$, we execute the integration by parts:

$$\int_{-\infty}^{+\infty} \varphi(q) (i\hbar \partial/\partial q) e^{-i(pq/\hbar)} dq = \int_{-\infty}^{+\infty} e^{-i(pq/\hbar)} (-i\hbar \partial/\partial q) \varphi(q) dq;$$

consequently,

$$\langle \chi|p|\chi \rangle = \int_{-\infty}^{+\infty} \varphi^*(q') dq' \int_{-\infty}^{+\infty} (-i\hbar \partial/\partial q) \varphi(q) dq \int_{-\infty}^{+\infty} e^{i(p(q'-q)/\hbar)} \frac{dp}{2\pi\hbar}.$$

As
$$\int_{-\infty}^{+\infty} e^{i(p(q'-q)/\hbar)} \frac{dp}{2\pi\hbar} = \delta(q'-q),$$

then

$$\langle \chi | p | \chi \rangle = \int_{-\infty}^{+\infty} (-i\hbar \partial/\partial q)\varphi(q) dq \int_{-\infty}^{+\infty} \varphi^*(q')\delta(q'-q) dq'$$

$$= \int_{-\infty}^{+\infty} \varphi^*(q)(-i\hbar \partial/\partial q)\varphi(q) dq.$$

Hence,

$$\langle \chi | p | \chi \rangle = \int_{-\infty}^{+\infty} \chi^*(p) p \chi(p) dp$$

$$= \int_{-\infty}^{+\infty} \varphi^*(q)(-i\hbar \partial/\partial q)\varphi(q) dq = \langle \varphi | p | \varphi \rangle$$

and, in the coordinate representation, the momentum is expressible as a linear differential operator:

$$p = -i\hbar \frac{\partial}{\partial q}.$$

With respect to coordinate, the analogous considerations are valid. In this case, we apply the inverse Fourier transformation

$$\varphi(q) = \frac{1}{\sqrt{2\pi\hbar}} \int_{-\infty}^{+\infty} \chi(p) e^{i(pq/\hbar)} dp,$$

p and q are simply interchanged and i is converted into $-i$. As a result,

$$q = i\hbar \frac{\partial}{\partial p}$$

represents the expression for the operator of coordinate in the momentum representation.

Having determined a commutator for two quantities G and F as follows,

$$[G, F] = GF - FG,$$

we see that

$$[q, -i\hbar \partial/\partial q]\varphi(q) = i\hbar \varphi(q) \text{ and } [i\hbar \partial/\partial p, p]\chi(p) = i\hbar \chi(p);$$

hence,

$$qp - pq = i\hbar,$$

and the conjugate coordinate and momentum in quantum mechanics fail to conform to the law of commutative multiplication. This condition constitutes properly the main distinction between quantum and classical theories. Classical physics operates with commuting to each other *c-numbers*, which are generally measurable in the experiment. The natural variables of quantum physics are opposite, *unobservable q-numbers*. The quantities of q-type, failing to commute generally with each other, act in a Hilbert space of abstract state vectors. The vectors are supposed to be transformed to each other through the action of q-numbers, which are in a sense the pertinent operators. Not all mathematical operators find their application in practice in physics. The principal requirements are *linearity* and *hermitivity*. Let

$$|n\rangle \text{ and } |m\rangle$$

be arbitrary vectors, and let

$$|\Psi\rangle = b|n\rangle + c|m\rangle,$$

in which b and c are numerical coefficients. By definition, G is the linear operator if

$$G|\Psi\rangle = bG|n\rangle + cG|m\rangle;$$

G is the Hermitian operator if

$$G = G^+,$$

in which G^+ satisfies the relation

$$\langle n|G|m\rangle = \langle m|G^+|n\rangle^*$$

and represents an operator Hermitian conjugate to G. For instance, as is easily seen in the coordinate representation, the differential operator for momentum possesses the properties of both hermitivity and linearity.

However, there exist such states in which the dynamic variables of quantum origin have determinate values belonging to a class of observable *c*-numbers. We imply here the *eigenvalues* of physical operators. So, if q-number G acts on state vector $|n\rangle$ without altering its "direction," an equation

$$G|n\rangle = g_n|n\rangle$$

is valid and g_n is a number of *c*-type, then $|n\rangle$ and g_n are, respectively, eigenvector and eigenvalue of operator G. In a general case, G has a set of eigenvectors, each

Ideas and principles

of which corresponds to a determinate eigenvalue. We show that the eigenvalues of physical operators are strictly real numbers. So, on the one side,

$$\langle n|G|n\rangle = g_n;$$

on the other side, taking into account that $G = G^+$, we have

$$\langle n|G|n\rangle = \langle n|G^+|n\rangle^* = \langle n|G|n\rangle^*;$$

consequently,

$$g_n = g_n^*$$

completely proves our assertion.

The eigenvectors that belong to various eigenvalues are mutually orthogonal. Let $|n\rangle$ and $|m\rangle$ be eigenvectors of physical operator G; they correspond to the equations for eigenvalues, respectively, g_n and g_m:

$$G|n\rangle = g_n|n\rangle \text{ and } G|m\rangle = g_m|m\rangle.$$

As $\langle m|G^+ = g_m^*\langle m|$, $G^+ = G$, and $g_m^* = g_m$, then

$$\langle m|G|n\rangle = g_n\langle m|n\rangle \text{ and } \langle m|G|n\rangle = g_m\langle m|n\rangle.$$

On comparing, we find

$$(g_n - g_m)\langle m|n\rangle = 0,$$

hence

$$\langle m|n\rangle = 0,$$

because $g_n \neq g_m$. The property of mutual orthogonality of eigenvectors plays an important role in the problems of quantum mechanics. Arbitrary vector $|\Psi\rangle$, for instance, can be represented in a form of expansion

$$|\Psi\rangle = \sum_n c_n|n\rangle,$$

in which c_n are the amplitudes of states in terms of the complete system of eigenvectors $|n\rangle$ of some operator G. If g_n are the eigenvalues corresponding to states n, then the average value of G in state Ψ is given by the expression

$$\langle \Psi|G|\Psi\rangle = \sum_{mn} c_m^* c_n \langle m|G|n\rangle = \sum_n |c_n|^2 g_n$$

that, in probability theory, represents a formula for an expectation value of quantity G.

Finally, we discuss a useful result regarding the possibility of simultaneous measurement of two dynamical quantities. Suppose that

$$G|n\rangle = g_n|n\rangle,$$

in which $|n\rangle$ and g_n represent, respectively, the eigenvectors and eigenvalues of physical operator G. On acting on this equation with some variable F on the left, we have

$$FG|n\rangle = g_n F|n\rangle.$$

If $[G, F] = 0$, then

$$G(F|n\rangle) = g_n(F|n\rangle);$$

consequently, quantity $F|n\rangle$ is the eigenvector of operator G. Hence,

$$F|n\rangle = f_n|n\rangle,$$

in which f_n are c-numbers. Commutative variables G and F thus have a common complete system of eigenstates. In this case, the dynamical quantities are simultaneously measurable.

Noncommutative physics

A *Hamiltonian* formalism underlies wave mechanics. Recall, in the classical theory, a motion of a system is described by a Hamiltonian H that is a function of canonically conjugate coordinates q_s and momenta p_s; index s numbers the degrees of freedom of the system. Quantity H represents a function of a total energy. Hamilton's equations of motion have the form

$$\frac{\partial H}{\partial p_s} = \dot{q}_s \text{ and } \frac{\partial H}{\partial q_s} = -\dot{p}_s,$$

in which a point above q_s and above p_s designates differentiation with respect to time t. In this case,

$$\frac{dG}{dt} = \sum_s \left(\frac{\partial G}{\partial q_s} \dot{q}_s + \frac{\partial G}{\partial p_s} \dot{p}_s \right) = \sum_s \left(\frac{\partial G}{\partial q_s} \frac{\partial H}{\partial p_s} - \frac{\partial G}{\partial p_s} \frac{\partial H}{\partial q_s} \right)$$

is the total derivative of dynamical variable G, which is a function of momenta and coordinates, with respect to time. If we introduce a Poisson *bracket*

$$\{F, I\} = \sum_s \left(\frac{\partial F}{\partial q_s} \frac{\partial I}{\partial p_s} - \frac{\partial F}{\partial p_s} \frac{\partial I}{\partial q_s} \right),$$

in which F and I are the arbitrary variables, then

$$\frac{dG}{dt} = \{G, H\}.$$

According to Dirac,[2] the transition into quantum mechanics is performed through a replacement of a Poisson bracket by a commutator divided by $i\hbar$, i.e.,

$$\{F, I\} \to \frac{1}{i\hbar}[F, I].$$

Let us note a series of simple properties

$$\{F, \text{const}\} = 0, \quad \{F, F\} = 0, \quad \{F, I\} = -\{I, F\},$$
$$\{F + G, I\} = \{F, I\} + \{G, I\}, \quad \text{and} \quad \{FG, I\} = \{F, I\}G + F\{G, I\},$$

which also hold equally for commutators, because

$$[F, \text{const}] = 0, \quad [F, F] = 0, \quad [F, I] = -[I, F],$$
$$[F + G, I] = [F, I] + [G, I], \quad \text{and} \quad [FG, I] = [F, I]G + F[G, I].$$

For canonically conjugate coordinates and momenta in classical mechanics, there exist the *fundamental* relations

$$\{q_s, q_k\} = 0, \{p_s, p_k\} = 0, \text{ and } \{q_s, p_k\} = \delta_{sk}.$$

Proceeding to the quantum quantities, we have

$$[q_s, q_k] = 0, \quad [p_s, p_k] = 0, \quad \text{and} \quad [q_s, p_k] = i\hbar\delta_{sk}.$$

We see that the value found for commutator,

$$[q_s, p_k],$$

agrees exactly with the result that has already been obtained previously for other reasons. For the *fundamental* commutators, equalities represent the *quantum conditions*, which the mutually conjugate coordinates and momenta must satisfy. One might, in an obvious manner, derive similar conditions for other couples of dynamical quantities, which are functions of the canonical variables. The quantum conditions cease to exist as $\hbar \to 0$; in this case, all variables become commutative and we proceed to theory classical from theory quantum.

Substituting now

$$\{G, H\} \to \frac{1}{i\hbar}[G, H],$$

we obtain an equation

$$i\hbar \frac{dG}{dt} = [G, H],$$

known as the *Heisenberg's equation of motion* that describes the *dynamical* variation of variable G of quantum origin with time. If G, in addition, depends explicitly on time, then the equation of motion should be written in a form

$$i\hbar \frac{dG}{dt} = i\hbar \frac{\partial G}{\partial t} + [G, H];$$

recall, in classical mechanics,

$$\frac{dG}{dt} = \frac{\partial G}{\partial t} + \{G, H\}.$$

The *constants of motion* play a special role. As a constant of motion, we understand some quantity G that satisfies the equation

$$\frac{dG}{dt} = 0.$$

In this case, with the proviso that

$$\frac{\partial G}{\partial t} = 0,$$

we have

$$[G, H] = 0$$

— the energy of a system and a dynamical variable, which represents a constant of motion, have *simultaneously determinate values*.

For instance, we consider a motion of a particle in a field with a potential V. Borrowing the Hamiltonian from classical theory, we suppose

$$H = \frac{\mathbf{p}^2}{2m} + V(\mathbf{r}),$$

in which m is the mass of the particle and $\mathbf{r}$ and $\mathbf{p}$ are its radius vector and momentum, respectively. According to Heisenberg's equation,

$$i\hbar \frac{d\mathbf{p}}{dt} = [\mathbf{p}, H] = [\mathbf{p}, V] = -i\hbar \frac{\partial V}{\partial \mathbf{r}},$$

because, in the representation of Cartesian coordinates, $\mathbf{r} = (x, y, z)$ and

$$\mathbf{p} = \left(-i\hbar\frac{\partial}{\partial x}, -i\hbar\frac{\partial}{\partial y}, -i\hbar\frac{\partial}{\partial z}\right) = -i\hbar\frac{\partial}{\partial \mathbf{r}}.$$

Thus, we eventually find

$$\frac{d\mathbf{p}}{dt} = -\frac{\partial V}{\partial \mathbf{r}}.$$

One might interpret this equation as a "quantum Newton's equation." The coincidence with the equation of motion of classical mechanics is, obviously, purely formal, because, in quantum theory, the dynamical variables do not have physical meaning, but their eigenvalues, which are substantially determined from other equations, do.

However, there exists another way to solve the problems of quantum mechanics. According to Schrödinger, the state vectors depend on time, but not on the dynamical variables. Furthermore, to represent the energy of a system, we attribute a differential operator

$$i\hbar\frac{\partial}{\partial t};$$

such a definition is consistent with the special theory of relativity because the relation between energy and time must be similar to the relation between coordinate and momentum. Thus, we should postulate the equation

$$i\hbar\frac{\partial \varphi}{\partial t} = H\varphi,$$

in which φ is the wavefunction of the system. This equation is called *Schrödinger's wave equation*; it describes the temporal variation of the wavefunction of the system. Working mainly in the coordinate representation, for Hamiltonian H of a particle moving in a field V, Schrödinger proposed the following linear operator:

$$H = -\frac{\hbar^2}{2m}\nabla^2 + V(\mathbf{r}, t),$$

in which m and $\mathbf{r}$ are the mass and radius vector of the particle, respectively. This operator, as is easily seen, is obtainable directly from classical expression

$$\frac{\mathbf{p}^2}{2m} + V(\mathbf{r}, t)$$

through the replacement of momentum $\mathbf{p}$ by operator

$$-i\hbar\nabla;$$

$\nabla = \partial/\partial \mathbf{r}$. We note that ∇^2 represents the *Laplace operator*; for instance,

$$\nabla^2 = \frac{1}{r^2}\frac{\partial}{\partial r}\left(r^2\frac{\partial}{\partial r}\right) + \frac{1}{r^2}\left[\frac{1}{\sin\theta}\frac{\partial}{\partial \theta}\left(\sin\theta\frac{\partial}{\partial \theta}\right) + \frac{1}{\sin^2\theta}\frac{\partial^2}{\partial \phi^2}\right]$$

is the expression for the Laplacian in spherical coordinates r, θ, and ϕ.

If the Hamiltonian has no explicit dependence on time, then the wave equation is simplified. In Schrödinger's equation,

$$\varphi = e^{-iEt/\hbar}\Phi \text{ and } \Phi \to |\Phi\rangle,$$

in which a function Φ is independent of time, we obtain

$$H|\Phi\rangle = E|\Phi\rangle;$$

here, one should understand H as the physical operator, whereas E is the c-number. This is Schrödinger's equation for the *stationary states* — for the states that possess a determinate value of energy. Possible values of energy E can be both discrete and continuous. From a mathematical point of view, E and Φ incarnate the corresponding eigenvalues and eigenfunctions of operator H. The set of stationary states is complete.

The names of Schrödinger and Heisenberg are strongly "linked" with each other; with them we associate the principal equations of quantum mechanics. Two pictures represent two points of view of one and the same. It seems these pictures are entirely equivalent. However, we should note that there exist such problems when the solutions, which are obtained in different pictures, become different. Here, we imply the problems of quantum electrodynamics, the consideration of which is yet to come.

Moment of momentum

In quantum theory, angular momentum $\mathbf{L}$ of a particle with radius vector $\mathbf{r}$ and momentum $\mathbf{p}$ are introduced through an expression

$$\mathbf{L} = \mathbf{r} \times \mathbf{p}$$

that, in a formal manner, corresponds exactly to the definition of $\mathbf{L}$ in classical mechanics. In Cartesian coordinates,

$$\mathbf{r} = (x, y, z) \text{ and } \mathbf{p} = (p_x, p_y, p_z);$$

therefore,

$$\mathbf{L} = (L_x, L_y, L_z) = (yp_z - zp_y, zp_x - xp_z, xp_y - yp_x).$$

There is no need to make concrete the order for various factors of projections of $\mathbf{r}$ and $\mathbf{p}$, because variable y commutes with p_z, z with p_y, x with p_z, and so on.

Cartesian components of vector operators **L** and **r**, and also **L** and **p**, generally fail to commute with each other. Indeed,

$$[L_x, x] = [yp_z - zp_y, x] = 0,$$
$$[L_x, y] = [yp_z - zp_y, y] = -z[p_y, y] = i\hbar z,$$

and

$$[L_x, z] = [yp_z - zp_y, z] = y[p_z, z] = -i\hbar y;$$

analogously,

$$[L_x, p_x] = [yp_z - zp_y, p_x] = 0,$$
$$[L_x, p_y] = [yp_z - zp_y, p_y] = [y, p_y]p_z = i\hbar p_z,$$

and

$$[L_x, p_z] = [yp_z - zp_y, p_z] = -[z, p_z]p_y = -i\hbar p_y.$$

Other relations are obtainable through a cyclic permutation of x, y, and z; for instance,

$$[L_y, z] = i\hbar x \rightarrow [L_z, x] = i\hbar y \rightarrow [L_x, y] = i\hbar z$$

and

$$[L_y, p_z] = i\hbar p_x \rightarrow [L_z, p_x] = i\hbar p_y \rightarrow [L_x, p_y] = i\hbar p_z.$$

The commutation relations for **L** and **r** are, hence, exactly analogous to those for **L** and **p**. If $\mathbf{a} = (a_x, a_y, a_z)$ is by definition **r** or **p**,

$$[L_x, a_x] = 0, \quad [L_x, a_y] = i\hbar a_z, \quad [L_x, a_z] = -i\hbar a_y, \quad \ldots$$

Let $\mathbf{b} = (b_x, b_y, b_z)$ also be **r** or **p**; then

$$[L_x, \mathbf{a} \cdot \mathbf{b}] = [L_x, a_x b_x + a_y b_y + a_z b_z]$$
$$= a_y[L_x, b_y] + [L_x, a_y]b_y + a_z[L_x, b_z] + [L_x, a_z]b_z = 0;$$

accordingly,

$$[L_y, \mathbf{a} \cdot \mathbf{b}] = 0 \quad \text{and} \quad [L_z, \mathbf{a} \cdot \mathbf{b}] = 0.$$

Any scalar consisting of **a** and **b** thus commutes with **L**:

$$[L_i, \mathbf{r}^2] = 0, \quad [L_i, \mathbf{p}^2] = 0, \quad [L_i, \mathbf{r} \cdot \mathbf{p}] = 0, \quad \ldots,$$

in which i denotes x or y or z.

We produce the commutation relations for the components of angular momentum **L**:

$$[L_x, L_y] = [L_x, zp_x - xp_z] = [L_x, z]p_x - x[L_x, p_z] = i\hbar(xp_y - yp_x) = i\hbar L_z;$$

through cyclic permutations, we obtain other commutators

$$[L_y, L_z] = i\hbar L_x \text{ and } [L_z, L_x] = i\hbar L_y$$

that are compactly expressible in a vector form

$$\mathbf{L} \times \mathbf{L} = i\hbar \mathbf{L}.$$

This formula is not quite absurd; one should bear in mind that components L_x, L_y, and L_z fail to commute with each other. With the aid of the commutation relations for moment of momentum, it is easy to prove that

$$[L_x, L_x^2] = 0,$$
$$[L_x, L_y^2] = L_y[L_x, L_y] + [L_x, L_y]L_y = i\hbar(L_y L_z + L_z L_y),$$

and

$$[L_x, L_z^2] = L_z[L_x, L_z] + [L_x, L_z]L_z = -i\hbar(L_z L_y + L_y L_z).$$

On summing these equalities, one finds

$$[L_x, \mathbf{L}^2] = 0,$$

in which

$$\mathbf{L}^2 = L_x^2 + L_y^2 + L_z^2.$$

In an analogous manner,

$$[L_y, \mathbf{L}^2] = 0 \text{ and } [L_z, \mathbf{L}^2] = 0.$$

The squared angular momentum, commuting with each component of vector **L**, thus might be simultaneously measured with one projection L_x or L_y or L_z. The projections of **L** fail to be commutative quantities with each other; therefore, they are not measurable in one state.

If projection L_z is defined, then instead of indeterminate quantities L_x and L_y, it is convenient to choose another pair of operators

$$L_+ = L_x + iL_y \text{ and } L_- = L_x - iL_y.$$

One accordingly performs the next relations:

$$[L_+, L_-] = -i[L_x, L_y] + i[L_y, L_x] = 2\hbar L_z,$$
$$[L_z, L_+] = [L_z, L_x] + i[L_z, L_y] = \hbar L_+,$$
$$[L_z, L_-] = [L_z, L_x] - i[L_z, L_y] = -\hbar L_-,$$
$$\mathbf{L}^2 = L_- L_+ + \hbar L_z + L_z^2 = L_+ L_- - \hbar L_z + L_z^2$$

plus the well-known expressions for differential operators of angular momentum in spherical coordinates r, θ, and ϕ:

$$L_\pm = \hbar e^{\pm i\phi} \left(\pm \frac{\partial}{\partial \theta} + i\,\mathrm{ctg}\,\theta \frac{\partial}{\partial \phi} \right),$$

$$L_z = -i\hbar \frac{\partial}{\partial \phi},$$

and

$$\mathbf{L}^2 = -\hbar^2 \left[\frac{1}{\sin^2\theta} \frac{\partial^2}{\partial \phi^2} + \frac{1}{\sin\theta} \frac{\partial}{\partial \theta}\left(\sin\theta \frac{\partial}{\partial \theta} \right) \right] \equiv -\hbar^2 \nabla^2_{\theta\phi},$$

in which $\nabla^2_{\theta\phi}$ is an angular part of the Laplace operator.

We calculate eigenvalues of operators L_z and $\mathbf{L}^2$; in this representation, L_x and L_y have indeterminate values. Let φ be eigenvectors and L'_z be eigenvalues of L_z; then,

$$-i\hbar \frac{\partial}{\partial \phi} \varphi(\phi) = L'_z \varphi(\phi).$$

This equation is readily integrated; as a result,

$$\varphi(\phi) = \frac{1}{\sqrt{2\pi}} C(r, \theta) e^{iL'_z \phi/\hbar}.$$

Function φ must be periodic in ϕ; the eigenvalues of projection L_z are consequently integral multiples of $\hbar$:

$$L'_z = \hbar k, \quad k = 0, \pm 1, \pm 2, \pm 3, \ldots$$

Here, $C(r, \theta)$ is a constant of integration; factor $1/\sqrt{2\pi}$ appears through a normalization condition

$$\frac{1}{2\pi} \int_0^{2\pi} e^{i(k'-k)\phi} d\phi = \delta_{k'k}.$$

If instead of L_z we choose, for instance, L_x, then we arrive at the same result, but just for the x-component of the angular momentum. In this representation, projections L_z and L_y then have no determinate value. An exception to this rule is the case

$$L_x = L_y = L_z = 0;$$

then $\mathbf{L}^2 = 0$ and all projections of $\mathbf{L}$ are simultaneously measurable.

We proceed to calculate eigenvalues of the squared angular momentum. As $L_z L_+ = L_+ L_z + \hbar L_+$, we have

$$L_z L_+ |\varphi_k\rangle = \hbar(k+1) L_+ |\varphi_k\rangle,$$

in which we took $L_z|\varphi_k\rangle = \hbar k |\varphi_k\rangle$ into account. Vector $L_+|\varphi_k\rangle$ is consequently the eigenvector of projection L_z belonging to eigenvalue $\hbar(k+1)$, which is accurate within a constant coefficient; hence,

$$|\varphi_{k+1}\rangle \sim L_+ |\varphi_k\rangle.$$

In an analogous manner, applying commutator $[L_z, L_-] = -\hbar L_-$, one might obtain that

$$|\varphi_{k-1}\rangle \sim L_- |\varphi_k\rangle.$$

Thus, L_+ is the operator that increases the value of k by unity and L_- is the operator that decreases k by unity.

We apply the nonnegativity of expression

$$\mathbf{L}^2 - L_z^2 = L_x^2 + L_y^2.$$

As

$$\mathbf{L}^2 - L_z^2$$

possesses only positive eigenvalues, there must exist an upper limit for L_z'; we denote it as $\hbar \ell$, in which ℓ is a positive integer. The states with $k > \ell$, by definition, do not exist; therefore, one must satisfy the equation $L_+|\varphi_\ell\rangle = 0$. On acting on this equality with the lowering operator on the left, one obtains

$$L_- L_+ |\varphi_\ell\rangle = (\mathbf{L}^2 - L_z^2 - \hbar L_z)|\varphi_\ell\rangle = 0.$$

Generally, $|\varphi_\ell\rangle \neq 0$; denoting the eigenvalue of $\mathbf{L}^2$ as Λ, we have

$$\Lambda - \hbar^2 \ell^2 - \hbar^2 \ell = 0,$$

therefore

$$\Lambda = \hbar^2 \ell(\ell+1).$$

Moreover, one should note these useful relations

$$L_\pm |\ell k\rangle = \hbar\sqrt{(\ell \mp k)(\ell \pm k + 1)}|\ell, k \pm 1\rangle,$$

which we implicitly applied and which are worthy of proof. We act in turn with raising and lowering operators on vector $|\varphi_k\rangle$, which is equal, by definition, to $|\ell k\rangle$; as a result,

$$L_-(L_+|\ell k\rangle) = \hbar^2(\ell - k)(\ell + k + 1)|\ell k\rangle = (\hbar^2\ell(\ell + 1) - \hbar^2 k^2 - \hbar^2 k)|\ell k\rangle$$
$$= (\mathbf{L}^2 - L_z^2 - \hbar L_z)|\ell k\rangle.$$

As $L_-L_+ = \mathbf{L}^2 - L_z^2 - \hbar L_z$, the above relations become proven.
Thus,

$$\mathbf{L}^2|\ell k\rangle = \hbar^2\ell(\ell + 1)|\ell k\rangle, \quad \ell = 0, 1, 2, \ldots$$

and

$$L_z|\ell k\rangle = \hbar k|\ell k\rangle, \quad k = 0, \pm 1, \ldots, \pm\ell.$$

Quantum number ℓ defines the squared angular momentum and might be equal to some nonnegative integer, including zero. Quantities of projection $\mathbf{L}$ along a selected direction are integral multiples of constant $\hbar$. For each ℓ, quantity $L_z'/\hbar$ runs over all negative and positive integers from $-\ell$ to $+\ell$. As a result, the state with a particular and nonzero number ℓ becomes degenerate. The degeneracy numbers $2\ell + 1$; hence, that many functions belong to eigenvalue $\hbar^2\ell(\ell + 1)$.

Perturbation theory

The main ideas of perturbation theory appeared in the second half of the eighteenth century; physicists were then involved in the problems of celestial mechanics. The purpose, as stated by Newton, was generally to investigate the interaction of celestial objects; in particular, Newton solved this problem for two bodies. Serious difficulties arose, however, in solving the many-body problem. In this connection, the approximate methods, among which the *method of perturbation theory* has achieved a special recognition, underwent rapid development. This approach involves the search for solutions in the form of infinite series.

A similar problem exists in quantum theory, because the quantum-mechanical equations possess exact solutions in only some trivial cases. In all other cases, one must apply approximate methods, to which, again, the perturbation method belongs. Restricting to a *stationary* theory, we consider this method in detail.

Let H_0 be an unperturbed Hamiltonian of a system. Under the influence of perturbation H', the Hamiltonian of the system alters, acquiring a form

$$H = H_0 + H',$$

in which

$$H' = \lambda W.$$

In essence, a factor λ represents a small parameter that characterizes the order of perturbation operator W. Our purpose is to determine the eigenvalues $E_n(\lambda)$ and eigenvectors $|\psi_n(\lambda)\rangle$ of perturbed Hamiltonian H, supposing that eigenvectors $|n\rangle$ and eigenvalues E_n^0 of Hamiltonian H_0 are known.

So, we differentiate Schrödinger's equation

$$(H_0 + \lambda W)|\psi_n(\lambda)\rangle = E_n(\lambda)|\psi_n(\lambda)\rangle$$

for stationary states with respect to λ; as a result,

$$(H_0 + \lambda W)\frac{d}{d\lambda}|\psi_n(\lambda)\rangle + W|\psi_n(\lambda)\rangle = \frac{dE_n(\lambda)}{d\lambda}|\psi_n(\lambda)\rangle + E_n(\lambda)\frac{d}{d\lambda}|\psi_n(\lambda)\rangle.$$

Applying the completeness and orthogonality of eigenfunctions and taking into account this equality,

$$\langle\psi_n(\lambda)|(d/d\lambda)|\psi_n(\lambda)\rangle = 0,$$

which follows from an obvious identity

$$\frac{d}{d\lambda}\langle\psi_n(\lambda)|\psi_n(\lambda)\rangle = \frac{d}{d\lambda}(1) = 0,$$

we represent $d|\psi_n(\lambda)\rangle/d\lambda$ as $\sum_{m\neq n} C_{mn}|\psi_m(\lambda)\rangle$; then,

$$\sum_{m\neq n} C_{mn}(E_n(\lambda) - E_m(\lambda))|\psi_m(\lambda)\rangle = W|\psi_n(\lambda)\rangle - \frac{dE_n(\lambda)}{d\lambda}|\psi_n(\lambda)\rangle.$$

One obtains immediately the system of equations[3]

$$\frac{d}{d\lambda}E_n(\lambda) = \langle\psi_n(\lambda)|W|\psi_n(\lambda)\rangle$$

and

$$\frac{d}{d\lambda}|\psi_n(\lambda)\rangle = \sum_{m\neq n} \frac{\langle\psi_m(\lambda)|W|\psi_n(\lambda)\rangle}{E_n(\lambda) - E_m(\lambda)}|\psi_m(\lambda)\rangle.$$

The summation is taken over all states of the perturbed Hamiltonian.

The obtained system is entirely equivalent to the initial Schrödinger equation and explicitly demonstrates that the calculations performed in terms of this perturbation theory have a recurrent character. Introducing into this system the sought corrections

$$E_n^s \text{ and } |\psi_n^s\rangle,$$

in which $s = 1, 2, \ldots$, through the series expansions

$$E_n(\lambda) = E_n^0 + \lambda E_n^1 + \lambda^2 E_n^2 + \cdots \text{ and } |\psi_n(\lambda)\rangle = |n\rangle + \lambda|\psi_n^1\rangle + \lambda^2|\psi_n^2\rangle + \cdots,$$

we compare quantities of the same order in parameter λ. We see that, through the first derivative with respect to λ, the corrections on the left side of the equations are 1 order of magnitude greater than that of the corrections on the right side. Hence,

$$E_n^1 = \langle n|W|n\rangle, \quad |\psi_n^1\rangle = \sum_{m \neq n} \frac{\langle m|W|n\rangle}{E_n^0 - E_m^0} |m\rangle \text{ and so on.}$$

As with a solution of any problem in a framework of perturbation theory, one must initially calculate the first-order corrections; this method becomes the most appropriate in many cases for those of higher order, if required, only afterwards.

Factorization

Stationary problems of quantum mechanics are generally considered within Schrödinger's picture. In this case, the principal equations are solved in the coordinate representation in a framework of the standard theory of integro-differential equations. Many problems might otherwise be solved in a purely algebraic manner. We focus our attention on the second possibility to seek a solution; we consider a simple and elegant algebraic method of finding the eigenvalues of physical operators that are explicitly independent of time. This method is factorization, described by Green.[4]

Suppose that there exist in a set the q-numbers

$$\eta_1, \eta_2, \text{ and so on.}$$

We determine an operator

$$F = \eta_1^+ \eta_1 + f_1 \equiv F_1,$$

in which f_1 is the physical number that has the maximum value possible for this representation. One uses, by definition,

$$F_2 = \eta_1 \eta_1^+ + f_1;$$

otherwise, we suggest that

$$F_2 = \eta_2^+ \eta_2 + f_2,$$

in which $f_2 \geq f_1$. According to this scenario, for arbitrary positive integer n, we have

$$F_n = \eta_n^+ \eta_n + f_n$$

and

$$F_{n+1} = \eta_n \eta_n^+ + f_n.$$

Let us introduce a vector

$$|\varphi_n\rangle = \eta_n \eta_{n-1} \cdots \eta_1 |\psi\rangle,$$

in which $|\psi\rangle$ is some normalized eigenvector of operator F belonging to eigenvalue f. As

$$F|\psi\rangle = f|\psi\rangle \text{ and } \langle\psi|\psi\rangle = 1,$$

then

$$\langle\varphi_1|\varphi_1\rangle = \langle\psi|\eta_1^+ \eta_1|\psi\rangle = \langle\psi|(F - f_1)|\psi\rangle = f - f_1;$$

taking into account that

$$\langle\varphi_1|\varphi_1\rangle \geq 0,$$

there follows this inequality,

$$f \geq f_1.$$

Furthermore,

$$F_{n+1}\eta_n = \eta_n \eta_n^+ \eta_n + \eta_n f_n = \eta_n F_n;$$

consequently,

$$\begin{aligned}
\langle\varphi_n|\varphi_n\rangle &= \langle\psi|\eta_1^+ \eta_2^+ \cdots \eta_n^+ \eta_n \eta_{n-1} \cdots \eta_1|\psi\rangle \\
&= \langle\psi|\eta_1^+ \eta_2^+ \cdots (F_n - f_n)\eta_{n-1} \cdots \eta_1|\psi\rangle \\
&= \langle\psi|\eta_1^+ \eta_2^+ \cdots \eta_{n-1}^+ \eta_{n-1}(F_{n-1} - f_n) \cdots \eta_1|\psi\rangle \\
&= \langle\psi|\eta_1^+ \eta_2^+ \cdots \eta_{n-1}^+ \eta_{n-1} \cdots \eta_1(F_1 - f_n)|\psi\rangle \\
&= \langle\psi|\eta_1^+ \eta_2^+ \cdots (F_{n-1} - f_{n-1})\eta_{n-2} \cdots \eta_1|\psi\rangle(f - f_n) \\
&= \cdots = (f - f_1)(f - f_2) \cdots (f - f_n) \geq 0.
\end{aligned}$$

Either

$$f \geq f_n$$

or

$$(f - f_1)(f - f_2) \cdots (f - f_{n-1}) = 0;$$

quantity f is thus either more than each physical number $f_1, f_2, \ldots, f_n$ or equal to one of them. The obtained result is highly important: quantities $f_1, f_2, \ldots$, which are represented in order of increasing magnitude, constitute the eigenvalues of operator F.

We proceed to construct the eigenvectors of operator F. Let $|n - 1\rangle$ be the eigenvector of F with eigenvalue f_n; one assumes that $|\psi\rangle = |n - 1\rangle$. We have

$$\langle \varphi_{n-1} | \varphi_{n-1} \rangle > 0$$

and

$$\langle \varphi_n | \varphi_n \rangle = \langle n - 1 | \eta_1^+ \eta_2^+ \ldots \eta_{n-1}^+ \eta_{n-1} \ldots \eta_1 (F - f_n) | n - 1 \rangle = 0,$$

because $F|n - 1\rangle = f_n |n - 1\rangle$; hence,

$$|\varphi_n\rangle = 0 \quad \text{or} \quad \eta_n | \varphi_{n-1} \rangle = 0.$$

Furthermore,

$$(F_n - f_n) | \varphi_{n-1} \rangle = \eta_n^+ \eta_n | \varphi_{n-1} \rangle = 0;$$

f_n is, hence, an eigenvalue of operator F_n with vector $|\varphi_{n-1}\rangle$. As

$$F_n \eta_n^+ = \eta_n^+ \eta_n \eta_n^+ + \eta_n^+ f_n = \eta_n^+ F_{n+1},$$

then multiplying $\eta_1^+ \eta_2^+ \ldots \eta_{n-1}^+$ by F_n, we obtain

$$\eta_1^+ \eta_2^+ \ldots \eta_{n-1}^+ F_n = \eta_1^+ \eta_2^+ \ldots F_{n-1} \eta_{n-1}^+ = F \eta_1^+ \eta_2^+ \ldots \eta_{n-1}^+$$

and

$$\eta_1^+ \eta_2^+ \ldots \eta_{n-1}^+ F_n | \varphi_{n-1} \rangle = F \eta_1^+ \eta_2^+ \ldots \eta_{n-1}^+ | \varphi_{n-1} \rangle = f_n \eta_1^+ \eta_2^+ \ldots \eta_{n-1}^+ | \varphi_{n-1} \rangle.$$

Operating on the other side, $F|n - 1\rangle = f_n|n - 1\rangle$. That result is consequently accurate within a constant factor,

$$|n - 1\rangle = \eta_1^+ \eta_2^+ \ldots \eta_{n-1}^+ | \varphi_{n-1} \rangle,$$

in which vector $|\varphi_{n-1}\rangle$ is determined by the equation

$$\eta_n | \varphi_{n-1} \rangle = 0;$$

a constant factor is chosen so that vector $|n - 1\rangle$ becomes normalized to unity.

Oscillator

As an example of the method just developed, we consider a harmonic oscillator. This example is of importance to understand physical processes, which are concerned with atomic and molecular vibrations, the theory of radiation, aspects of quantum field theory, and many other questions. The Hamiltonian of a one-dimensional harmonic oscillator is given in the form

$$H = \frac{p^2}{2m} + \frac{m\omega^2 x^2}{2},$$

in which appear mass m, momentum p, and displacement x from an equilibrium point of a particle that makes small vibrations with frequency ω; quantities x and p satisfy the commutation relation

$[x, p] = i\hbar.$

Supposing x and p to be classical variables, we transform Hamiltonian H. We have

$$H_{\text{class}} = \hbar\omega\left(\frac{p^2}{2m\omega\hbar} + \frac{m\omega}{2\hbar}x^2\right) = \hbar\omega\left(\sqrt{\frac{m\omega}{2\hbar}}x - i\frac{p}{\sqrt{2m\omega\hbar}}\right)\left(\sqrt{\frac{m\omega}{2\hbar}}x + i\frac{p}{\sqrt{2m\omega\hbar}}\right),$$

in which one should understand H_{class} in a classical meaning such that

$xp = px.$

We introduce a new quantity,

$$\eta = \frac{1}{\sqrt{2}}\left(\sqrt{\frac{m\omega}{\hbar}}x + i\frac{p}{\sqrt{m\omega\hbar}}\right);$$

then,

$H_{\text{class}} = \hbar\omega\eta^*\eta.$

Let us seek what this classical expression yields in quantum mechanics. Supposing x and p to now be operators, we assume

$$\eta = \frac{1}{\sqrt{2}}\left(\sqrt{\frac{m\omega}{\hbar}}x + i\frac{p}{\sqrt{m\omega\hbar}}\right) \text{ and } \eta^+ = \frac{1}{\sqrt{2}}\left(\sqrt{\frac{m\omega}{\hbar}}x - i\frac{p}{\sqrt{m\omega\hbar}}\right);$$

in this case,

$$[\eta, \eta^+] = -\frac{i}{\hbar}[x, p] = 1.$$

Here, we take into account that $x = x^+$ and $p = p^+$. Furthermore,

$H_{\text{class}} \rightarrow \hbar\omega\eta^+\eta;$

consequently,

$$\hbar\omega\eta^+\eta = \frac{p^2}{2m} + \frac{m\omega^2 x^2}{2} + \frac{i}{2\hbar}\hbar\omega[x,p] = H - \frac{\hbar\omega}{2},$$

hence

$$H = \hbar\omega\left(\eta^+\eta + \frac{1}{2}\right).$$

The distinction between H and H_{class} consists of the appearance of an additional constant quantity $\hbar\omega/2$.

To find the energy levels of a harmonic oscillator, one must therefore solve the problem for the eigenvalues of operator

$$F = \eta^+\eta + \frac{1}{2}$$

with the condition that

$$[\eta, \eta^+] = 1.$$

We apply the method of factorization. For all n, by definition, we assume

$$\eta_n = \eta.$$

Because

$$F = F_1 = \eta_1^+\eta_1 + f_1 \text{ and } \eta_1 = \eta,$$

then

$$f_1 = \frac{1}{2};$$

for an oscillator, this value corresponds to the level with the least energy

$$E_0 = \hbar\omega f_1 = \frac{\hbar\omega}{2}.$$

We find other eigenvalues from a comparison of the two expressions for F_{n+1}. We have

$$\eta_n\eta_n^+ + f_n = \eta_{n+1}^+\eta_{n+1} + f_{n+1};$$

hence,

$$f_{n+1} = f_n + [\eta, \eta^+] = f_n + 1 = f_{n-1} + 2 = \cdots = f_1 + n.$$

To quantity f_{n+1} corresponds the value of energy

$$E_n = \hbar\omega f_{n+1},$$

because $H = \hbar\omega F$ and

$$H|n\rangle = E_n|n\rangle,$$

in which $|n\rangle$ is a respective eigenvector of the Hamiltonian. Consequently,

$$E_n = \hbar\omega\left(n + \frac{1}{2}\right), \quad n = 0, 1, 2, \ldots,$$

here, we slightly redefined number n, having included the value $n = 0$. The sought expression for the energy of a harmonic oscillator is found. The adjacent *quantized levels* are separate from each other by constant quantity $\hbar\omega$, such that the levels of the oscillator are distributed in an equidistant manner. The least possible value of energy equals

$$\hbar\omega/2,$$

not zero, as in classical mechanics.

Let us construct the system of eigenvectors $|n\rangle$ for the values of energy E_n of the oscillator. So,

$$|n\rangle = \eta_1^+ \eta_2^+ \ldots \eta_n^+ |\varphi_n\rangle,$$

in which $\eta_{n+1}|\varphi_n\rangle = 0$. As $\eta_{n+1} = \eta$, then for all n it is convenient to use

$$|\varphi_n\rangle = |0\rangle.$$

Vector $|0\rangle$, incarnating the ground state vector of the oscillator, is the solution of this equation

$$\eta|0\rangle = 0;$$

by definition,

$$\langle 0|0\rangle = 1.$$

Taking into account the equality $F_n \eta_n^+ = \eta_n^+ F_{n+1}$, we normalize vector $|n\rangle$; we have

$$\langle m|n\rangle = \langle\varphi_m|\eta_m\ldots\eta_2\eta_1\eta_1^+\eta_2^+\ldots\eta_n^+|\varphi_n\rangle$$
$$= \langle\varphi_m|\eta_m\ldots\eta_2(F_2-f_1)\eta_2^+\ldots\eta_n^+|\varphi_n\rangle$$
$$= \langle\varphi_m|\eta_m\ldots\eta_2\eta_2^+\ldots\eta_n^+(F_{n+1}-f_1)|\varphi_n\rangle$$
$$= \langle\varphi_m|\eta_m\ldots\eta_3(F_3-f_2)\eta_3^+\ldots\eta_n^+|\varphi_n\rangle(f_{n+1}-f_1)$$
$$= \cdots = \langle\varphi_m|\eta_m\ldots\eta_{n+1}|\varphi_n\rangle(f_{n+1}-f_1)(f_{n+1}-f_2)\ldots(f_{n+1}-f_n)$$
$$= \langle 0|\eta^{m-n}|0\rangle n! = n!\delta_{mn},$$

in which, for a determinacy, $m \geq n$. Consequently, assuming that a normalization factor is equal to $(n!)^{-1/2}$, we obtain

$$|n\rangle = \frac{1}{\sqrt{n!}}(\eta^+)^n|0\rangle.$$

In such a form, vectors $|n\rangle$ form the sought orthonormal system of eigenvectors of a harmonic oscillator.

We note also these useful relations

$$\eta^+\eta|n\rangle = n|n\rangle, \quad \eta^+|n-1\rangle = \sqrt{n}|n\rangle, \quad \text{and} \quad \eta|n\rangle = \sqrt{n}|n-1\rangle,$$

which are easily proven. Indeed, as $F - f_1 = \eta^+\eta$, then

$$\eta^+\eta|n\rangle = (f_{n+1}-f_1)|n\rangle = n|n\rangle;$$

furthermore,

$$|n\rangle = \frac{1}{\sqrt{n!}}(\eta^+)^n|0\rangle = \frac{\eta^+}{\sqrt{n}}\frac{1}{\sqrt{(n-1)!}}(\eta^+)^{n-1}|0\rangle = \frac{\eta^+}{\sqrt{n}}|n-1\rangle$$

and, finally,

$$\eta|n\rangle = \frac{1}{\sqrt{n}}\eta\eta^+|n-1\rangle = \frac{1}{\sqrt{n}}(\eta^+\eta+1)|n-1\rangle = \sqrt{n}|n-1\rangle.$$

Quantum numbers

Let us consider the important generalizations of the developed theory of an oscillator. We begin with a choice of Hamiltonians. The simplest choice to describe the vibrations of an arbitrary system in quantum mechanics is a model of harmonic oscillators. In this case, the Hamiltonian has a form

$$H^0 = \hbar\sum_{k=1}^{r}\frac{\omega_k}{2}(p_k^2 + q_k^2),$$

in which ω_k are the harmonic frequencies, r is the total number of *normal* vibrations, and p_k and q_k are the momenta and their conjugate coordinates; note that quantities p_k and q_k are dimensionless here. The energy levels of a system are defined as a sum of the energies of separate oscillators and are expressible as

$$E_n^0 = \hbar \sum_k \omega_k \left(n_k + \frac{1}{2}\right), \quad n_k = 0, 1, 2, \ldots$$

Eigenvectors

$$|n_1, n_2, \ldots, n_k, \ldots\rangle$$

of Hamiltonian H^0, corresponding to eigenvalues E_n^0, represent the products of individual vectors $|n_1\rangle, |n_2\rangle, \ldots$ of harmonic one-dimensional oscillators.

A harmonic case is certainly only an idealization of vibrations of a real system. The potential energy V of vibrations, as in classical physics, is generally written in a form of an expansion in terms of normal coordinates q_k:

$$V = V_0 + \frac{1}{2}\hbar \sum_k \omega_k q_k^2 + \sum_{ijk} a_{ijk}^{(1)} q_i q_j q_k + \sum_{ijk\ell} a_{ijk\ell}^{(2)} q_i q_j q_k q_\ell + \cdots,$$

in which the linear terms disappear through the fact that the first derivative V' equals zero at the equilibrium condition; variables q_k are chosen so that in V the terms of type $q_k q_\ell$ with $k \neq \ell$ disappear; V_0 is the minimum of function V. We see that a set of harmonic oscillators corresponds to a first approximation. This model is only qualitatively correct; in fact, vibrations, failing to conform to a harmonic law, are anharmonic. To describe correctly the vibrations, apart from the quadratic part of the potential energy, one must therefore take into account the normal coordinates to greater than quadratic powers in an expansion of V. These terms additional to H^0 are defined by anharmonicity coefficients

$$a_{ijk}^{(1)}, a_{ijk\ell}^{(2)}, \ldots$$

and characterize the interactions among various vibrational modes. The calculation of the corresponding corrections is generally performed with a perturbation theory for stationary states of a Hamiltonian of a general type

$$H = H^0 + W,$$

in which perturbation function represents an expansion in powers of a small parameter λ:

$$W = \sum_{p>0} \lambda^p \sum_{(j_1 j_2 \ldots j_r) p+2} a_{j_1 j_2 \ldots j_r} \xi_1^{j_1} \xi_2^{j_2} \ldots \xi_r^{j_r} \equiv \sum_{p>0} \lambda^p G_p,$$

$$\xi_k = \sqrt{2} q_k, \quad k = 1, 2, \ldots, r;$$

$a_{j_1j_2...j_r}$ are the anharmonic force coefficients. A special summation is performed over the indices in parentheses, symbol $(j_1j_2...j_r)p+2$ signifies a summation over $j_1,j_2,...,j_r$ under the constraint that

$$j_1+j_2+\cdots+j_r=p+2.$$

In the following, when using such a summation, we denote the set of indices associated with the normal variables, e.g., $j_1,j_2,...$ and j_r as j.

The eigenvalues of Hamiltonian H, incarnating the *quantized* energy levels of the anharmonic oscillator, can be represented in a form of the expansion

$$E_n = \sum_{i_1i_2...i_r} \Omega_{i_1i_2...i_r}\left(n_1+\frac{1}{2}\right)^{i_1}\left(n_2+\frac{1}{2}\right)^{i_2}\cdots\left(n_r+\frac{1}{2}\right)^{i_r}$$

in terms of quantum numbers; coefficients $\Omega_{i_1i_2...i_r}$ are expressible through $a_{j_1j_2...j_r}$ and ω_k with the aid of the formulae of perturbation theory. Expansion E_n is a somewhat particular case of this more general expression,[5]

$$(n|F|n+k) = \sqrt{g_{n_1,n_1+k_1}\cdots g_{n_r,n_r+k_r}}\sum_{i_1...i_r}\Phi^{i_1...i_r}_{k_1...k_r}\left(n_1+\frac{k_1+1}{2}\right)^{i_1}\cdots\left(n_r+\frac{k_r+1}{2}\right)^{i_r}$$

for the matrix element of some physical operator F between eigenvectors $|n)$ and $|n+k)$ of Hamiltonian H, because

$$E_n = (n|H|n);$$

$\Phi^{i_1...i_r}_{k_1...k_r}$ are coefficients,

$$g_{n_\ell,n_\ell+k_\ell} = (n_\ell+1)(n_\ell+2)...(n_\ell+k_\ell) = \frac{(n_\ell+k_\ell)!}{n_\ell!},$$

numbers k_ℓ, in which $\ell = 1, 2, ...$ and r, are nonnegative integers, including zero. These relations are proved in a framework of the *formalism of polynomials of quantum numbers* in Chapter 3. Note that E_n is generally a function of quantum numbers $n_1+1/2$, $n_2+1/2$, ..., $n_r+1/2$. Moreover, E_n depends additionally on some parameters that determine the influence of anharmonicity. Varying these parameters and an explicit form of function E_n, we substantially obtain various representations of anharmonicity. Analogous considerations are valid for function Φ that depends on quantum numbers $n_1+k_1/2+1/2$, $n_2+k_2/2+1/2$, ..., $n_r+k_r/2+1/2$, and determines an explicit form of arbitrary matrix element

$$(n|F|n+k) = \sqrt{g}\Phi,$$

in which $g = g_{n_1,n_1+k_1}g_{n_2,n_2+k_2}\cdots g_{n_r,n_r+k_r}$.

Physics of the electron

Hydrogen atom

Considering an atom or a molecule, we imply some stationary state of all electrons moving in a field of atomic nuclei. Because the problem is stationary, it is most convenient to solve it according to Schrödinger's picture. For atoms, the problem is simplified because, in this case, one might assume that every electron is in an effective centrally symmetric field of a nucleus and other electrons. This case is highly important; we consider in detail the features of the motion of an electron in a central force field.

Let us begin with a general discussion. For an electron of mass m, in the representation of spherical coordinates r, θ, and ϕ, we write Schrödinger's equation

$$-\frac{\hbar^2}{2m}\left[\frac{1}{r^2}\frac{\partial}{\partial r}\left(r^2\frac{\partial}{\partial r}\right) + \frac{1}{r^2}\nabla^2_{\theta\phi}\right]|\Phi\rangle + V(r)|\Phi\rangle = E|\Phi\rangle$$

for stationary states Φ. Here, $V(r)$ is the potential of a field having spherical symmetry and E is the energy of the electron. Taking into account that squared angular momentum $\mathbf{L}^2$ is directly linked with an angular part $\nabla^2_{\theta\phi}$ of the Laplace operator through this relation

$$\mathbf{L}^2 = -\hbar^2\nabla^2_{\theta\phi},$$

we rewrite our equation as

$$\left[-\frac{\hbar^2}{2m}\frac{1}{r^2}\frac{\partial}{\partial r}\left(r^2\frac{\partial}{\partial r}\right) + \frac{\mathbf{L}^2}{2mr^2} + V(r)\right]|\Phi\rangle = E|\Phi\rangle.$$

Operator $\mathbf{L}^2$ acts on only angular variables, not on r; consequently, one might separate variables as

$$|\Phi\rangle = Q(r)|\ell k\rangle,$$

in which $|\ell k\rangle$ are eigenvectors of squared angular momentum that are characterized with quantum numbers ℓ and k. As

$$\mathbf{L}^2|\ell k\rangle = \hbar^2\ell(\ell+1)|\ell k\rangle,$$

a radial wave function $Q(r)$ satisfies the equation

$$-\frac{\hbar^2}{2m}\frac{1}{r^2}\frac{\partial}{\partial r}\left(r^2\frac{\partial Q}{\partial r}\right) + \frac{\hbar^2\ell(\ell+1)}{2mr^2}Q + V(r)Q = EQ.$$

This is a radial Schrödinger equation. A simple substitution

$$Q(r) = \frac{\varsigma(r)}{r}$$

yields the equation

$$H\varsigma = E\varsigma$$

for eigenvalues E and eigenfunctions ς of this Hamiltonian,

$$H = \frac{p_r^2}{2m} + \frac{\hbar^2\ell(\ell+1)}{2mr^2} + V(r),$$

in which

$$p_r = -i\hbar\frac{\partial}{\partial r}.$$

As a result, we obtain a one-dimensional problem for the motion of the electron in an effective field with the potential

$$V(r) + \frac{\hbar^2\ell(\ell+1)}{2mr^2};$$

the additional term, arising from $\mathbf{L}^2/2mr^2$, represents substantially centrifugal energy.

Note that

$$|\Phi|^2 r^2 \sin\theta dr d\theta d\phi$$

is the probability to find an electron in an element of volume $r^2 \sin\theta dr d\theta d\phi$. Having integrated with respect to the angles, we find that the probability of finding of an electron in a spherical layer of thickness dr is $|Q|^2 r^2 dr$ or $|\varsigma|^2 dr$. Because $r \in [0, \infty)$, it is necessary that integral

$$\int_0^\infty |\varsigma|^2 dr$$

converges at $r = 0$, which is at the lower limit. For this proviso, it is sufficient to use $\varsigma(0) = 0$. This is the first boundary condition for the radial wave function.

Regarding a second boundary condition, because $r \to \infty$, a wave function fails to tend to infinity, although it can have a finite nonzero value.

Bohr's formula

As an example of applying the radial Schrödinger's equation, we consider the hydrogen atom and determine the energy levels of an electron in a Coulombic field. In this case,

$$V(r) = -\frac{e^2}{r}$$

and

$$H = \frac{p_r^2}{2m} + \frac{\hbar^2 \ell(\ell+1)}{2mr^2} - \frac{e^2}{r},$$

in which r is the distance between the electron of charge $-e$ and the nucleus of charge $+e$. To solve this problem, following Green,[4] we use the method of factorization. We assume

$$F = 2mH,$$

that is,

$$F = p_r^2 + \frac{\hbar^2 \ell(\ell+1)}{r^2} - \frac{2\kappa}{r}, \quad \kappa = e^2 m.$$

Let

$$\eta_n = p_r + i\left(a_n + \frac{b_n}{r}\right),$$

in which a_n and b_n are real quantities, the explicit form of which one must find. We calculate $\eta_n^+ \eta_n$ as

$$\eta_n^+ \eta_n = (p_r - i(a_n + b_n/r))(p_r + i(a_n + b_n/r))$$
$$= p_r^2 + ib_n[p_r, 1/r] + a_n^2 + \frac{2a_n b_n}{r} + \frac{b_n^2}{r^2} = p_r^2 + a_n^2 + \frac{2a_n b_n}{r} + \frac{b_n^2 - \hbar b_n}{r^2},$$

in which one takes into account that

$$[p_r, 1/r] = -i\hbar \frac{\partial}{\partial r}(1/r) = \frac{i\hbar}{r^2}.$$

In an analogous manner, one obtains

$$\eta_n \eta_n^+ = p_r^2 + a_n^2 + \frac{2a_n b_n}{r} + \frac{b_n^2 + \hbar b_n}{r^2}.$$

We define operator F_1:

$$F_1 = \eta_1^+ \eta_1 + f_1 = p_r^2 + a_1^2 + \frac{2a_1 b_1}{r} + \frac{b_1^2 - \hbar b_1}{r^2} + f_1.$$

On the other side,

$$F_1 = p_r^2 + \frac{\hbar^2 \ell(\ell+1)}{r^2} - \frac{2\kappa}{r}.$$

In comparison, we obtain the equations

$$a_1 b_1 = -\kappa, \quad b_1(b_1 - \hbar) = \hbar^2 \ell(\ell+1) \text{ and } a_1^2 + f_1 = 0.$$

Here, one might find two solutions. In the first case,

$$b_1 = -\hbar \ell,$$

then

$$a_1 = \frac{\kappa}{\hbar \ell} \text{ and } f_1 = -\frac{\kappa^2}{\hbar^2 \ell^2}.$$

If

$$b_1 = \hbar(\ell+1),$$

then

$$a_1 = -\frac{\kappa}{\hbar(\ell+1)} \text{ and } f_1 = -\frac{\kappa^2}{\hbar^2(\ell+1)^2}.$$

Because

$$-\frac{\kappa^2}{\hbar^2(\ell+1)^2} > -\frac{\kappa^2}{\hbar^2 \ell^2},$$

one chooses the second case.

We compare the two expressions for F_{n+1}:

$$\eta_n \eta_n^+ + f_n = \eta_{n+1}^+ \eta_{n+1} + f_{n+1}$$

or, in an explicit form,

$$p_r^2 + a_n^2 + \frac{2a_n b_n}{r} + \frac{b_n^2 + \hbar b_n}{r^2} + f_n = p_r^2 + a_{n+1}^2 + \frac{2a_{n+1} b_{n+1}}{r} + \frac{b_{n+1}^2 - \hbar b_{n+1}}{r^2} + f_{n+1},$$

hence

$$a_n b_n = a_{n+1} b_{n+1},$$

$$b_n(b_n + \hbar) = b_{n+1}(b_{n+1} - \hbar),$$

and

$$a_n^2 + f_n = a_{n+1}^2 + f_{n+1}.$$

If

$$b_n = -b_{n+1}, \text{ then } a_n = -a_{n+1} \text{ and } f_n = f_{n+1};$$

this case fails to hold physical interest.

Let us consider the second possibility, for which

$$b_{n+1} = b_n + \hbar.$$

We have

$$b_{n+1} = b_n + \hbar = b_{n-1} + 2\hbar = \cdots = b_1 + n\hbar = \hbar(\ell + n + 1);$$

$$a_{n+1} b_{n+1} = a_n b_n = a_{n-1} b_{n-1} = \cdots = a_1 b_1 = -\kappa,$$

therefore

$$a_n = -\frac{\kappa}{b_n} = -\frac{\kappa}{\hbar(\ell + n)}.$$

Taking into account that

$$a_{n+1}^2 + f_{n+1} = a_n^2 + f_n = \cdots = a_1^2 + f_1 = 0,$$

we find the eigenvalues of quantity F:

$$f_n = -a_n^2 = -\frac{\kappa^2}{\hbar^2(\ell+n)^2}.$$

Eigenvalues E of Hamiltonian H are thus

$$E = \frac{f_n}{2m} = -\frac{\kappa^2}{2m\hbar^2(\ell+n)^2}$$

or

$$E_\nu = -\frac{me^4}{2\hbar^2\nu^2}, \quad \nu = 1, 2, \ldots$$

Because the set of c-numbers E_1, E_2, and so on, is restricted by value $E = 0$, then, according to this inequality

$$(E - E_1)(E - E_2)\ldots(E - E_n) \geq 0$$

of Green's formalism, quantity E must be either equal to one value E_1, E_2, ... or equal to any value from zero until infinity. For $E < 0$, the energy levels constitute a discrete spectrum for which the electron is in a bound state. For $E > 0$, there is no bound state of the electron; the energy spectrum is continuous.

Because $\nu \to \infty$, the discrete spectrum, to which the levels with negative values of energy correspond, is converted to continuous through the value $E = 0$. Number ν is called a principal quantum number. It is important that the energy depends on only ν, not on quantum numbers ℓ and k, which characterize the angular momentum. Each eigenvalue E_ν is therefore degenerate. Taking into account that

$$\nu \geq \ell + 1,$$

for given ν, ℓ fails to exceed value $\nu - 1$; consequently, the total degeneracy in a Coulombic field equals

$$\sum_{\ell=0}^{\nu-1}(2\ell+1) = \nu^2,$$

that is, to each value E_ν, ν^2 various states correspond.

The obtained expression for E_ν of hydrogen atom agrees perfectly with Bohr's famous formula for frequencies

$$\omega_{\nu'\nu} = \frac{E_{\nu'} - E_\nu}{\hbar} = \frac{me^4}{2\hbar^3}\left(\frac{1}{\nu^2} - \frac{1}{\nu'^2}\right), \quad \nu' > \nu;$$

Bohr arrived at this formula from disparate considerations; quantity

$$me^4/2\hbar^3$$

substantially represents the Rydberg constant. The frequencies that belong to the transitions of the atom to a given level form a so-called spectral series. For instance, the transitions with frequencies

$$\omega_{\nu'1} = \frac{me^4}{2\hbar^3}\left(1 - \frac{1}{\nu'^2}\right)$$

to level $\nu = 1$ correspond to the Lyman series, the spectral lines of which lie in the vacuum-ultraviolet part of the optical spectrum. If $\nu = 2$, then the lines with frequencies $\omega_{\nu'2}$ in the visible spectrum represent the Balmer series. The transitions to level $\nu = 3$ form the Paschen–Ritz series; frequencies $\omega_{\nu'3}$ are displayed in the infrared spectrum.

Matrix elements

Let us obtain a general expression for the radial matrix elements of a hydrogen atom. It is convenient to work in the coordinate representation. In this case,

$$K_{n'n} = \int_0^\infty Q_{n'}^*(r)K(r)Q_n(r)r^2 dr = \int_0^\infty \varsigma_{n'}^*(r)K(r)\varsigma_n(r)dr$$

is the sought matrix element of arbitrary function $K(r)$; $\varsigma_n(r)$ is the normalized vector $\langle r|n-1\rangle$ that, in terms of operators for *creation* η_n^+ and *destruction* η_n, is given by the formula

$$\langle r|n-1\rangle = \eta_1^+ \eta_2^+ \ldots \eta_{n-1}^+ \langle r|\varphi_{n-1}\rangle,$$

in which $\langle r|\varphi_{n-1}\rangle$ or simply $\varphi_{n-1}(r)$ is defined from the equation

$$\eta_n \varphi_{n-1}(r) = 0.$$

We initially calculate $\varphi_{n-1}(r)$. Because

$$\eta_n = -i\hbar\frac{\partial}{\partial r} + i\left(a_n + \frac{b_n}{r}\right),$$

we have

$$\hbar\frac{\partial \varphi_{n-1}}{\partial r} = \left(a_n + \frac{b_n}{r}\right)\varphi_{n-1},$$

hence

$$\varphi_{n-1}(r) = \varphi_{n-1}^0 \cdot r^{b_n/\hbar} e^{a_n r/\hbar} = \varphi_{n-1}^0 \cdot r^{\ell+n} e^{-r\kappa(\ell+n)^{-1}\hbar^{-2}},$$

in which φ_{n-1}^0 is a constant of integration.

Therefore, we find $\langle r|n-1\rangle$,

$$\langle r|n-1\rangle = (-\mathrm{i})^{n-1}\left(\hbar\frac{\partial}{\partial r} + a_1 + \frac{b_1}{r}\right)\left(\hbar\frac{\partial}{\partial r} + a_2 + \frac{b_2}{r}\right)\ldots\left(\hbar\frac{\partial}{\partial r} + a_{n-1} + \frac{b_{n-1}}{r}\right)\varphi_{n-1},$$

in which we take into account that

$$\eta_n^+ = -\mathrm{i}\hbar\frac{\partial}{\partial r} - \mathrm{i}\left(a_n + \frac{b_n}{r}\right).$$

Thus,

$$\varsigma_n(r) = C \cdot \left(\hbar\frac{\partial}{\partial r} + a_1 + \frac{b_1}{r}\right)\left(\hbar\frac{\partial}{\partial r} + a_2 + \frac{b_2}{r}\right)\ldots\left(\hbar\frac{\partial}{\partial r} + a_{n-1} + \frac{b_{n-1}}{r}\right) r^{\ell+n} e^{-r\kappa(\ell+n)^{-1}\hbar^{-2}},$$

in which C is chosen so that

$$\int_0^\infty \varsigma_n^*(r)\varsigma_n(r)\mathrm{d}r = 1.$$

Quantities $\varsigma_n(r)$ are generally found; with their aid, for an explicitly specified function $K(r)$, one might calculate an arbitrary matrix element

$$K_{n'n} = \int_0^\infty \varsigma_{n'}^*(r) K(r) \varsigma_n(r)\mathrm{d}r.$$

In particular, the obtained expression for $K_{n'n}$ is applicable for calculating the observable intensities of transitions of the hydrogen atom.

Dirac's equation

Despite the success of Schrödinger's nonrelativistic theory, it is physically unsatisfactory. This theory fails to explain the spin of the electron, to yield the correct expression for fine structure, and to take into account the specification of quantum-electrodynamic effects. According to Dirac, the principal problem of the *old* theory involves how to choose a Hamiltonian H. In a nonrelativistic case,

$$H = \frac{\mathbf{p}^2}{2m} + \cdots,$$

in which m is the mass of a particle and $\mathbf{p}$ is its momentum; in Schrödinger's equation, there is no symmetry between space coordinates and time t, between quantity of energy

$$E \to i\hbar \frac{\partial}{\partial t}$$

and components of momentum p_x, p_y, and p_z. To increase the attraction to quantum theory, one should either combine Schrödinger's equation with a relativistic Hamiltonian or discover absolutely another Hamiltonian.

We consider the former scheme. Momentum $\mathbf{p}$ and energy E of a particle are related to each other, forming a four-vector

$$p^\mu = \left(\frac{E}{c}, p_x, p_y, p_z\right), \quad \mu = 0, 1, 2, 3,$$

such that

$$\sum_{\mu,\nu} g^{\mu\nu} p_\mu p_\nu = \sum_{\mu,\nu} g_{\mu\nu} p^\mu p^\nu \equiv p_\mu p^\mu = \left(\frac{E}{c}\right)^2 - \mathbf{p}^2 = m^2 c^2,$$

in which c is the speed of light, and

$$g^{\mu\nu} = \begin{pmatrix} 1 & 0 & 0 & 0 \\ 0 & -1 & 0 & 0 \\ 0 & 0 & -1 & 0 \\ 0 & 0 & 0 & -1 \end{pmatrix}$$

is Minkowski's metric tensor. In the classical expression

$$(E/c)^2 - \mathbf{p}^2 = m^2 c^2,$$

replacing E, according to Schrödinger's equation, with operator $i\hbar\partial/\partial t$, and $\mathbf{p}$ with operator $-i\hbar\nabla$, we obtain the equation

$$\left(\hbar^2 \frac{\partial^2}{\partial t^2} - \hbar^2 c^2 \nabla^2 + m^2 c^4\right)\psi = 0$$

or

$$(p_\mu p^\mu - m^2 c^2)\psi = 0,$$

in which ψ is the wave function of the particle, $p^\mu = i\hbar\partial/\partial x_\mu$ and $x_\mu = (ct, -\mathbf{r})$;

$$x^\mu = \sum_\nu g^{\mu\nu} x_\nu = (ct, \mathbf{r}) \text{ and } p_\mu = \sum_\nu g_{\mu\nu} p^\nu = (p_0, -\mathbf{p}).$$

Quantity

$$p_0 = p^0 = i\hbar \frac{\partial}{\partial x_0} = \frac{i\hbar}{c} \frac{\partial}{\partial t} \rightarrow \frac{E}{c}$$

represents a fourth temporal component of the momentum operator.

This first scheme to construct the relativistic quantum theory fails to become sufficiently informative; it yields a solution with a negative value of energy,

$$E = \pm\sqrt{\mathbf{p}^2 c^2 + m^2 c^4}.$$

Whether this situation is unsatisfactory becomes clear when, together with particles, antiparticles come under consideration. The obtained equation, which bears the names of Klein, Fock, and Gordon, is relativistically invariant and applicable to describe a particle with spin that equals zero. Schrödinger also obtained this equation.

Following Dirac, we consider the second scheme to modify Schrödinger's equation, which amounts to a search for a Hamiltonian of a new type. Substituting $i\hbar\partial/\partial t$ by cp_0, we have

$$p_0 \psi = \frac{H}{c} \psi \equiv (?)\psi.$$

As p_0 enters into the equation linearly, one expects other components of the four-vector of momentum to appear in the equation in a linear manner. Hence,

$$(?)\psi = (\alpha_1 p_1 + \alpha_2 p_2 + \alpha_3 p_3 + \beta)\psi,$$

such that

$$(p_0 - \alpha_1 p_1 - \alpha_2 p_2 - \alpha_3 p_3 - \beta)\psi = 0,$$

in which quantities α and β are independent of neither coordinates nor momenta; they describe the new degrees of freedom that are hidden from classical mechanics. We multiply this equation by $(p_0 + \alpha_1 p_1 + \alpha_2 p_2 + \alpha_3 p_3 + \beta)$ on the left,

$$\left(p_0^2 - \sum_r \alpha_r^2 p_r^2 - \beta^2 - \sum_{r \neq s} (\alpha_r \alpha_s + \alpha_s \alpha_r) p_r p_s - \sum_r (\alpha_r \beta + \beta \alpha_r) p_r \right) \psi = 0;$$

to bring the latter into coincidence with equation $(p_0^2 - \mathbf{p}^2 - m^2 c^2)\psi = 0$, one must assume

$$\alpha_r^2 = 1, \quad \beta^2 = m^2 c^2,$$

Physics of the electron

$$\alpha_r \alpha_s + \alpha_s \alpha_r = 0 \quad \text{at } r \neq s,$$

$$\alpha_r \beta + \beta \alpha_r = 0.$$

These relations are the equivalent of the well-known rules for Pauli matrices,

$$\sigma_r \sigma_s + \sigma_s \sigma_r = 2\delta_{rs},$$

in which

$$\sigma_1 = \begin{pmatrix} 0 & 1 \\ 1 & 0 \end{pmatrix}, \quad \sigma_2 = \begin{pmatrix} 0 & -i \\ i & 0 \end{pmatrix}, \quad \sigma_3 = \begin{pmatrix} 1 & 0 \\ 0 & -1 \end{pmatrix}.$$

We must have, however, four matrices, not three; 2×2 matrices are therefore insufficient for our purpose. Let us determine the minimum size N of the new matrices. For instance, $\alpha_1 \alpha_2 = -\alpha_2 \alpha_1$,

$$\det \alpha_1 \cdot \det \alpha_2 = \det(-I) \cdot \det \alpha_2 \cdot \det \alpha_1,$$

therefore

$$\det(-I) = (-1)^N = 1.$$

Number N is thus even and equal to at least four. Moreover, $\alpha_2 = -\alpha_1^{-1} \alpha_2 \alpha_1$, such that

$$\mathrm{Sp}(\alpha_2) = -\mathrm{Sp}(\alpha_1^{-1} \alpha_2 \alpha_1) = -\mathrm{Sp}(\alpha_2) = 0;$$

spurs $\mathrm{Sp}(\alpha_1)$, $\mathrm{Sp}(\alpha_3)$, and $\mathrm{Sp}(\beta)$ also equal zero.

To satisfy all these relations with regard to α and β, we extend the system of Pauli matrices in a diagonal manner

$$\sigma \rightarrow \begin{pmatrix} \sigma & 0 \\ 0 & \sigma \end{pmatrix},$$

that is,

$$\sigma_1 = \begin{pmatrix} 0 & 1 & 0 & 0 \\ 1 & 0 & 0 & 0 \\ 0 & 0 & 0 & 1 \\ 0 & 0 & 1 & 0 \end{pmatrix}, \quad \sigma_2 = \begin{pmatrix} 0 & -i & 0 & 0 \\ i & 0 & 0 & 0 \\ 0 & 0 & 0 & -i \\ 0 & 0 & i & 0 \end{pmatrix}, \quad \sigma_3 = \begin{pmatrix} 1 & 0 & 0 & 0 \\ 0 & -1 & 0 & 0 \\ 0 & 0 & 1 & 0 \\ 0 & 0 & 0 & -1 \end{pmatrix}.$$

We introduce three more matrices ρ_1, ρ_2, and ρ_3, having interchanged in σ_r the second and third rows and columns:

$$\rho = \left\{ \begin{pmatrix} 0 & I \\ I & 0 \end{pmatrix}, i\begin{pmatrix} 0 & -I \\ I & 0 \end{pmatrix}, \begin{pmatrix} I & 0 \\ 0 & -I \end{pmatrix} \right\},$$

in which I is a 2×2 unit matrix. As we see, ρ has the structure of Pauli matrices with 2×2 elements; obviously,

$$\rho_r \rho_s + \rho_s \rho_r = 2\delta_{rs} \text{ and } \rho_r \sigma_s = \sigma_s \rho_r.$$

According to Dirac, we assume

$$\alpha_r = \rho_1 \sigma_r \text{ and } \beta = mc\rho_3;$$

accordingly, with this definition,

$$\alpha_r^2 = \rho_1^2 \sigma_r^2 = 1,$$

$$\alpha_1 \alpha_2 = \rho_1^2 \sigma_1 \sigma_2 = -\rho_1^2 \sigma_2 \sigma_1 = -\alpha_2 \alpha_1,$$

and so on.
As a result,

$$(p_0 - \rho_1(\boldsymbol{\sigma} \cdot \mathbf{p}) - mc\rho_3)\psi = 0.$$

This equation, first derived by Dirac, describes particles with spin equal to one half. To rewrite the new equation in a covariant manner, we multiply it by ρ_3,

$$(\rho_3 p_0 - \rho_3 \rho_1(\boldsymbol{\sigma} \cdot \mathbf{p}) - mc)\psi = 0,$$

put, by definition, $\gamma^0 = \rho_3$ and $\gamma^r = \rho_3 \rho_1 \sigma_r$; consequently,

$$(\gamma^\mu p_\mu - mc)\psi = 0.$$

Latin indices correspond to three vectors, and Greek indices correspond to four vectors. Dirac's matrices have these explicit forms,

$$\gamma^0 = \begin{pmatrix} I & 0 \\ 0 & -I \end{pmatrix} \text{ and } \boldsymbol{\gamma} = \begin{pmatrix} 0 & \boldsymbol{\sigma} \\ -\boldsymbol{\sigma} & 0 \end{pmatrix},$$

in which, in a determination of $\boldsymbol{\gamma}$, 2×2 Pauli matrices appear,

$$\sigma_1 = \begin{pmatrix} 0 & 1 \\ 1 & 0 \end{pmatrix}, \quad \sigma_2 = \begin{pmatrix} 0 & -i \\ i & 0 \end{pmatrix}, \text{ and } \sigma_3 = \begin{pmatrix} 1 & 0 \\ 0 & -1 \end{pmatrix},$$

as the components of vector $\boldsymbol{\sigma}$; hereafter, in a pertinent context, we stipulate that, regarding σ_1, σ_2, and σ_3, one should understand quantities 2×2, not 4×4.

Relativistic invariance

In modern quantum theory, together with Schrödinger's and Heisenberg's pictures, Dirac's equation has a place similar to those of Lagrange's equations in mechanics and Maxwell's equations in electrodynamics. In the new wave equation, the relativistic structure and the rules of noncommutative algebra are naturally combined. There is no problem concerned with the negativity of the density of states, and the principal results are experimentally confirmed. However, the conformation to the theory of relativity demands additional elucidation. The new theory must yield results that are independent of the choice of a Lorentz frame of reference.

We consider a linear transformation from x^ν to x'^μ:

$$x^\nu = a^{\nu\mu} x'_\mu, \quad x'^\mu = a^{\mu\nu} x_\nu, \quad a_{\mu\nu} a^{\nu\tau} = \delta^\tau_\mu.$$

Suppose that Dirac's equation written in the new coordinates retains its initial form; that is,

$$\left(i\hbar \gamma^\mu \frac{\partial}{\partial x'^\mu} - mc \right) \psi' = 0,$$

in which ψ' is a function of coordinates x'^μ. With the aid of this transformation,

$$\psi' = S\psi,$$

we return to the initial variables. We have

$$\frac{\partial}{\partial x'^\mu} = \frac{\partial x^\nu}{\partial x'^\mu} \cdot \frac{\partial}{\partial x^\nu} = a^\nu_\mu \cdot \frac{\partial}{\partial x^\nu},$$

such that

$$\left(i\hbar a^\nu_\mu \gamma^\mu S \frac{\partial}{\partial x^\nu} - Smc \right) \psi = 0.$$

Through an orthogonality of transformation,

$$S^{-1} S = 1,$$

multiplying the obtained equation by S^{-1} on the left side, we consequently find

$$\left(i\hbar(S^{-1}a_\mu^\nu\gamma^\mu S)\frac{\partial}{\partial x^\nu} - mc\right)\psi = 0.$$

For this equation to coincide with Dirac's equation written with primed coordinates, one must enforce the equality

$$S^{-1}a_\mu^\nu\gamma^\mu S = \gamma^\nu$$

or

$$S\gamma^\nu S^{-1} = a_\mu^\nu\gamma^\mu.$$

To prove the Lorentz invariance, we must answer two questions. Is there a transformation S that preserves the form of the initial Dirac equation? Might matrix S imply a Lorentz transformation matrix?

We initially reply to the first query. We consider a linear rotational transformation, for instance, in plane x_1x_2. An expression for the rotation matrix is given in a form

$$a_\mu^\nu = \begin{pmatrix} 1 & 0 & 0 & 0 \\ 0 & \cos\phi & \sin\phi & 0 \\ 0 & -\sin\phi & \cos\phi & 0 \\ 0 & 0 & 0 & 1 \end{pmatrix}, \quad \begin{cases} x_1' = x_1\cos\phi + x_2\sin\phi, \\ x_2' = -x_1\sin\phi + x_2\cos\phi. \end{cases}$$

To show that $S = \exp(\phi\gamma^1\gamma^2/2)$, we have

$$S = 1 + \frac{\phi}{2}\gamma^1\gamma^2 + \frac{\phi^2}{2!\cdot 4}(\gamma^1\gamma^2)^2 + \frac{\phi^3}{3!\cdot 8}(\gamma^1\gamma^2)^3 + \frac{\phi^4}{4!\cdot 16}(\gamma^1\gamma^2)^4 + \cdots;$$

as

$$(\gamma^1\gamma^2)^2 = \gamma^1\gamma^2\gamma^1\gamma^2 = -(\gamma^1)^2(\gamma^2)^2 = -1,$$

$$(\gamma^1\gamma^2)^3 = (\gamma^1\gamma^2)^2\gamma^1\gamma^2 = -\gamma^1\gamma^2,$$

$$(\gamma^1\gamma^2)^4 = +1,$$

and so on,

$$S = \left(1 - \frac{\phi^2}{2!\cdot 4} + \frac{\phi^4}{4!\cdot 16} - \cdots\right) + \gamma^1\gamma^2\left(\frac{\phi}{2} - \frac{\phi^3}{3!\cdot 8} + \cdots\right) = \cos\frac{\phi}{2} + \gamma^1\gamma^2\sin\frac{\phi}{2}.$$

One readily verifies that $S^{-1}S = 1$ if

$$S^{-1} = \exp(-\phi\gamma^1\gamma^2/2) = \cos\frac{\phi}{2} - \gamma^1\gamma^2\sin\frac{\phi}{2}.$$

Finally,

$$S\gamma^\nu S^{-1} = \gamma^\nu\cos^2\left(\frac{\phi}{2}\right) - \gamma^\nu\gamma^1\gamma^2\cos\left(\frac{\phi}{2}\right)\sin\left(\frac{\phi}{2}\right)$$
$$+ \gamma^1\gamma^2\gamma^\nu\cos\left(\frac{\phi}{2}\right)\sin\left(\frac{\phi}{2}\right) - \gamma^1\gamma^2\gamma^\nu\gamma^1\gamma^2\sin^2\left(\frac{\phi}{2}\right),$$

therefore

$$S\gamma^1 S^{-1} = \gamma^1 \cos\phi + \gamma^2 \sin\phi, \quad S\gamma^2 S^{-1} = -\gamma^1 \sin\phi + \gamma^2 \cos\phi,$$
and $S\gamma^\nu S^{-1} = \gamma^\nu$ at $\nu \neq 1, 2$.

One sees that $S\gamma^\nu S^{-1} = a^\nu_\mu \gamma^\mu$, so that Dirac's equation is indeed invariant with regard to a rotational transformation.

There is then no major difficulty to answer the second question and to prove Lorentz invariance. A Lorentz transformation — a conversion to the system moving with regard to an initial system with velocity $v = $ const — is well known to represent a rotation in plane $x_1 x_0$ by an imaginary angle. Using $\phi = i\vartheta$ and bearing in mind the imaginary unit at the temporal coordinate x_0, we have

$$a^\nu_\mu = \begin{pmatrix} \text{ch}\vartheta & -\text{sh}\vartheta & 0 & 0 \\ -\text{sh}\vartheta & \text{ch}\vartheta & 0 & 0 \\ 0 & 0 & 1 & 0 \\ 0 & 0 & 0 & 1 \end{pmatrix}, \quad \begin{cases} x'_0 = x_0\text{ch}\vartheta - x_1\text{sh}\vartheta, & \text{th}\vartheta = v/c, \\ x'_1 = -x_0\text{sh}\vartheta + x_1\text{ch}\vartheta, & \text{ch}\vartheta = (1-(v/c)^2)^{-1/2}. \end{cases}$$

The sought transformation matrix acquires a form

$$S = \exp(i\vartheta\gamma^0\gamma^1/2) = \text{ch}(\vartheta/2) + i\gamma^0\gamma^1\text{sh}(\vartheta/2);$$

obviously,

$$S^{-1} = \exp(-i\vartheta\gamma^0\gamma^1/2) = \text{ch}(\vartheta/2) - i\gamma^0\gamma^1\text{sh}(\vartheta/2).$$

The wave equation of Dirac is thus relativistically invariant. It yields physical results that are independent of the Lorentz frame of reference.

Spin one-half

The wave equation of Dirac is essential to explain the doubling of stationary levels for an electron in an atom. According to Schrödinger's picture, one might circumvent this difficulty through a phenomenological introduction of an electron spin that equals $\hbar/2$ and a magnetic moment equal to the Bohr magneton $\mu_B = e\hbar/2mc$. Although Pauli, having heuristically applied this approach, succeeded in treating the new phenomenon, the nature of the pertinent degrees of freedom remained unclear. In this sense, Dirac's theory elucidated all aspects in question. Apart from an experimental confirmation, the spin and magnetic moment of the electron have acquired a solid theoretical foundation.

We extrapolate Dirac's equation to the case of the presence of an external electromagnetic field. As in classical physics, one should replace energy E by $E + eA_0$ and momentum $\mathbf{p}$ by $\mathbf{p} + e\mathbf{A}/c$, in which e is the absolute value of an electronic charge, c is the speed of light, and A_0 and $\mathbf{A}$ are corresponding scalar and vector potentials of a field. If $A^\mu = (A^0, \mathbf{A})$ is the four-vector of a field potential, then

$$p_\mu \rightarrow p_\mu + \frac{e}{c} A_\mu.$$

This replacement possesses both gradient and Lorentz invariance. As a result, we obtain

$$\left(\gamma^\mu \left(p_\mu + \frac{e}{c} A_\mu\right) - mc\right)\psi = 0,$$

in which m is the mass of electron. We multiply the obtained equation by $\gamma^\nu(p_\nu + eA_\nu/c)$ on the left to yield

$$\left(\gamma^\nu \gamma^\mu \left(p_\nu + \frac{e}{c} A_\nu\right)\left(p_\mu + \frac{e}{c} A_\mu\right) - m^2 c^2\right)\psi = 0.$$

One sees that γ-matrices of Dirac satisfy the relation of Clifford algebra

$$\gamma_\mu \gamma_\nu + \gamma_\nu \gamma_\mu = 2g_{\mu\nu}.$$

If this relation is combined with an antisymmetric tensor

$$\sigma^{\nu\mu} = -\sigma^{\mu\nu} = \frac{i}{2}(\gamma^\nu \gamma^\mu - \gamma^\mu \gamma^\nu),$$

then one might, directly, express the product $\gamma^\nu \gamma^\mu$ through $g^{\nu\mu}$ and $\sigma^{\nu\mu}$:

$$\gamma^\nu \gamma^\mu = g^{\nu\mu} - i\sigma^{\nu\mu}.$$

Consequently,

$$\gamma^\nu\gamma^\mu\left(p_\nu + \frac{eA_\nu}{c}\right)\left(p_\mu + \frac{eA_\mu}{c}\right)$$

$$= \left(p^\mu + \frac{eA^\mu}{c}\right)\left(p_\mu + \frac{eA_\mu}{c}\right) - i\sigma^{\nu\mu}\left(p_\nu + \frac{eA_\nu}{c}\right)\left(p_\mu + \frac{eA_\mu}{c}\right)$$

$$= \left(p^\mu + \frac{eA^\mu}{c}\right)\left(p_\mu + \frac{eA_\mu}{c}\right) - \frac{i}{2}(\sigma^{\nu\mu} - \sigma^{\mu\nu})\left(p_\nu + \frac{eA_\nu}{c}\right)\left(p_\mu + \frac{eA_\mu}{c}\right)$$

$$= \left(p^\mu + \frac{eA^\mu}{c}\right)\left(p_\mu + \frac{eA_\mu}{c}\right) - \frac{i}{2}\sigma^{\nu\mu}\left[p_\nu + \frac{eA_\nu}{c}, p_\mu + \frac{eA_\mu}{c}\right].$$

Here, for the commutator, we have

$$\left[p_\nu + \frac{eA_\nu}{c}, p_\mu + \frac{eA_\mu}{c}\right] = \left[p_\nu, \frac{eA_\mu}{c}\right] - \left[p_\mu, \frac{eA_\nu}{c}\right] = \frac{ie\hbar}{c}\left(\frac{\partial A_\mu}{\partial x^\nu} - \frac{\partial A_\nu}{\partial x^\mu}\right) = \frac{ie\hbar}{c}F_{\nu\mu},$$

in which

$$F_{\nu\mu} = \frac{\partial A_\mu}{\partial x^\nu} - \frac{\partial A_\nu}{\partial x^\mu} = \begin{pmatrix} 0 & R_1 & R_2 & R_3 \\ -R_1 & 0 & -B_3 & B_2 \\ -R_2 & B_3 & 0 & -B_1 \\ -R_3 & -B_2 & B_1 & 0 \end{pmatrix}$$

is the tensor of the electromagnetic field with a polar electric field vector $\mathbf{R} = (R_1, R_2, R_3)$ and an axial magnetic field vector $\mathbf{B} = (B_1, B_2, B_3)$.

The quadratic Dirac's equation in the external field acquires a form

$$\left(\left(p^\mu + \frac{e}{c}A^\mu\right)\left(p_\mu + \frac{e}{c}A_\mu\right) + \frac{e\hbar}{2c}\sigma^{\nu\mu}F_{\nu\mu} - m^2c^2\right)\psi = 0.$$

To simplify it, we use $\gamma_r = \rho_3\rho_1\sigma_r = i\rho_2\sigma_r$ and $\gamma_0 = \rho_3$. We have

$$\sigma^{0r} = \frac{1}{2}\sigma^r(\rho_2\rho_3 - \rho_3\rho_2) = i\rho_1\sigma^r = i\alpha^r,$$

$$\sigma^{12} = -\frac{i}{2}(\sigma^1\sigma^2 - \sigma^2\sigma^1) = -i\sigma^1\sigma^2 = \sigma^3, \sigma^{23} = \sigma^1, \sigma^{31} = \sigma^2.$$

Consequently,

$$\sigma^{\nu\mu}F_{\nu\mu} = 2\sigma^{0r}R_r + 2\sigma^{rs}F_{rs}|_{r<s} = 2i\alpha^r R_r - 2\sigma^r B_r,$$

and the equation for an electron in an external field becomes

$$\left(\left(\frac{E+eA_0}{c}\right)^2 - \left(\mathbf{p}+\frac{e\mathbf{A}}{c}\right)^2 - \frac{e\hbar}{c}(\boldsymbol{\sigma}\cdot\mathbf{B}) + i\frac{e\hbar}{c}(\boldsymbol{\alpha}\cdot\mathbf{R}) - m^2 c^2\right)\psi = 0.$$

Here, we perceive two supplementary terms

$$-\frac{e\hbar}{c}(\boldsymbol{\sigma}\cdot\mathbf{B}) \text{ and } i\frac{e\hbar}{c}(\boldsymbol{\alpha}\cdot\mathbf{R}).$$

The former shows the presence of the new degree of freedom for an electron—spin and the magnetic moment concerned with spin

$$\boldsymbol{\mu} = -\frac{e\hbar\boldsymbol{\sigma}}{2mc}$$

that interacts with an external magnetic field **B**. Spin emphasizes that an electron, possessing an inner mechanical angular momentum, "rotates" about its own axis. The latter term turns out to be imaginary; its principal purpose is to ensure the relativistic invariance of Dirac's theory.

According to a physical point of view, the purely imaginary term is of only minor interest, because it corresponds to a presence of an imaginary electric moment for the electron. One might suppose that its appearance is necessary only for that purpose, in a formal manner, to adapt the new theory to Schrödinger's picture. The latter is essentially nonrelativistic, and its role is therefore highly doubtful. Omitting this imaginary term, we define this nonrelativistic limit. One should assume that

$$E = \varepsilon + mc^2, \quad eA_0 \ll mc^2 \text{ and } \varepsilon \ll mc^2;$$

then,

$$\left(\frac{E+eA_0}{c}\right)^2 - m^2 c^2 \approx 2m(\varepsilon + eA_0)$$

and

$$\left(\frac{1}{2m}\left(\mathbf{p}+\frac{e\mathbf{A}}{c}\right)^2 - eA_0 + \frac{e\hbar}{2mc}(\boldsymbol{\sigma}\cdot\mathbf{B})\right)\psi = \varepsilon\psi$$

that constitutes the famous Pauli equation.

Applying another consideration, one might arrive at a definition of spin. The angular momentum in a central field of force, for which $\mathbf{A} = 0$ and $A_0 = A_0(r)$, is invariant. In this case, Dirac's Hamiltonian, additionally multiplied by c, has a form

$$H = -eA_0(r) + c\rho_1(\boldsymbol{\sigma} \cdot \mathbf{p}) + \rho_3 mc^2.$$

We calculate commutator $[\mathbf{L}, H]$, in which $\mathbf{L} = (L_1, L_2, L_3)$ is the orbital angular momentum of the electron. We have

$$[L_1, H] = c\rho_1 \boldsymbol{\sigma} \cdot [L_1, \mathbf{p}] = c\rho_1 \boldsymbol{\sigma} \cdot (\mathbf{j}[L_1, p_2] + \mathbf{k}[L_1, p_3])$$
$$= i\hbar c\rho_1(\sigma_2 p_3 - \sigma_3 p_2) = i\hbar c\rho_1(\boldsymbol{\sigma} \times \mathbf{p})_1;$$

consequently, $[\mathbf{L}, H] = i\hbar c\rho_1(\boldsymbol{\sigma} \times \mathbf{p})$, and angular momentum $\mathbf{L}$ fails to be invariant. We proceed to calculate $[\boldsymbol{\sigma}, H]$:

$$[\sigma_1, H] = c\rho_1[\sigma_1, \boldsymbol{\sigma}] \cdot \mathbf{p} = c\rho_1(\mathbf{j}[\sigma_1, \sigma_2] + \mathbf{k}[\sigma_1, \sigma_3]) \cdot \mathbf{p}$$
$$= 2ic\rho_1(\sigma_3 p_2 - \sigma_2 p_3) = -2ic\rho_1(\boldsymbol{\sigma} \times \mathbf{p})_1;$$

thus, $[\hbar\boldsymbol{\sigma}/2, H] = -i\hbar c\rho_1(\boldsymbol{\sigma} \times \mathbf{p})$. One sees that

$$\left[\mathbf{L} + \frac{\hbar}{2}\boldsymbol{\sigma}, H\right] = 0,$$

such that vector $\mathbf{L} + \hbar\boldsymbol{\sigma}/2$ is a constant of the motion. The electron thus possesses an inner angular momentum $\hbar\boldsymbol{\sigma}/2$, which is appropriately called spin. The eigenvalues of one projection of quantity $\boldsymbol{\sigma}$ equal ± 1, which conform entirely to the hypothesis of Goudsmit and Uhlenbeck; the observable values of spin momentum are $\pm\hbar/2$. Spin is an exceptional quantum quantity that tends to zero in a classical limit as $\hbar \to 0$.

The Pauli equation derived here is the result of a particular nonrelativistic limit for Dirac's theory. However, Dirac's equation admits another cardinal nonrelativistic consideration that yields physically correct results with no additional supposition, unlike what Pauli's phenomenological theory includes. To investigate this limiting case, we write Dirac's equation in an external electric field with potential A_0:

$$\left(\left(p_0 + \frac{eA_0}{c}\right) - \rho_1(\boldsymbol{\sigma} \cdot \mathbf{p}) - \rho_3 mc\right)\psi = 0,$$

or in an explicit form after multiplying by c:

$$\left((E + eA_0)\begin{pmatrix} I & 0 \\ 0 & I \end{pmatrix} - c(\boldsymbol{\sigma} \cdot \mathbf{p})\begin{pmatrix} 0 & I \\ I & 0 \end{pmatrix} - mc^2\begin{pmatrix} I & 0 \\ 0 & -I \end{pmatrix}\right)\begin{pmatrix} \psi_A \\ \psi_B \end{pmatrix} = 0.$$

Here,

$$\psi_A = \begin{pmatrix} \psi_a \\ \psi_{a'} \end{pmatrix} \text{ and } \psi_B = \begin{pmatrix} \psi_b \\ \psi_{b'} \end{pmatrix}$$

are two-component wave functions. This equation is equivalent to a system

$$c(\boldsymbol{\sigma} \cdot \mathbf{p})\psi_B + mc^2\psi_A = (E + eA_0)\psi_A,$$
$$c(\boldsymbol{\sigma} \cdot \mathbf{p})\psi_A - mc^2\psi_B = (E + eA_0)\psi_B,$$

in which, regarding $\boldsymbol{\sigma}$, we understand the vector, the components of which are 2×2 Pauli matrices. We assume $\varepsilon = E - mc^2$, isolate ψ_B from the latter equation, and substitute it into the former to yield

$$\psi_B = c(\varepsilon + eA_0 + 2mc^2)^{-1}(\boldsymbol{\sigma} \cdot \mathbf{p})\psi_A$$

and

$$\left(\frac{1}{2m}(\boldsymbol{\sigma} \cdot \mathbf{p})\left(1 + \frac{\varepsilon + eA_0}{2mc^2}\right)^{-1}(\boldsymbol{\sigma} \cdot \mathbf{p}) - eA_0 \right)\psi_A = \varepsilon\psi_A.$$

In the nonrelativistic case,

$$\mathbf{p} = m\mathbf{v}, \quad \varepsilon \ll mc^2 \text{ and } eA_0 \ll mc^2,$$

in which $\mathbf{v}$ is the velocity of the electron, such that

$$\psi_B \sim \frac{v}{c}|\boldsymbol{\sigma}|\psi_A$$

and two components

$$\psi_B = \begin{pmatrix} \psi_b \\ \psi_{b'} \end{pmatrix}$$

are appropriately called small. To define the large components ψ_A, we use the approximation

$$\left(1 + \frac{\varepsilon + eA_0}{2mc^2}\right)^{-1} \approx 1 - \frac{\varepsilon + eA_0}{2mc^2},$$

take into account that

$$\mathbf{p}A_0 = A_0\mathbf{p} - i\hbar\frac{\partial A_0}{\partial \mathbf{r}},$$

and notice equalities

$$(\boldsymbol{\sigma}\cdot\mathbf{p})^2 = \mathbf{p}^2$$

and

$$\left(\boldsymbol{\sigma}\cdot\frac{\partial A_0}{\partial \mathbf{r}}\right)(\boldsymbol{\sigma}\cdot\mathbf{p}) = \frac{\partial A_0}{\partial \mathbf{r}}\cdot\mathbf{p} + i\boldsymbol{\sigma}\left(\frac{\partial A_0}{\partial \mathbf{r}}\times\mathbf{p}\right),$$

which follow from the well-known relation

$$(\boldsymbol{\sigma}\cdot\mathbf{a})(\boldsymbol{\sigma}\cdot\mathbf{b}) = \mathbf{a}\cdot\mathbf{b} + i\boldsymbol{\sigma}\cdot(\mathbf{a}\times\mathbf{b}),$$

which is satisfied for arbitrary vectors **a** and **b** as a pair. Consequently,

$$(\boldsymbol{\sigma}\cdot\mathbf{p})A_0(\boldsymbol{\sigma}\cdot\mathbf{p}) = A_0\mathbf{p}^2 - i\hbar\left(\frac{\partial A_0}{\partial \mathbf{r}}\mathbf{p} + i\boldsymbol{\sigma}\left(\frac{\partial A_0}{\partial \mathbf{r}}\times\mathbf{p}\right)\right).$$

Supposing spherical symmetry for potential A_0, we have

$$\frac{\partial A_0}{\partial \mathbf{r}} = A_0'\frac{\mathbf{r}}{r}.$$

Thus,

$$\left(\frac{\mathbf{p}^2}{2m} - \left(\frac{\varepsilon + eA_0}{2mc^2}\right)\frac{\mathbf{p}^2}{2m} - eA_0 + \frac{ie\hbar}{4m^2c^2}\left(A_0'\frac{\mathbf{r}\cdot\mathbf{p}}{r} + iA_0'\boldsymbol{\sigma}\cdot\frac{\mathbf{r}\times\mathbf{p}}{r}\right)\right)\psi_A = \varepsilon\psi_A.$$

Noticing that

$$\frac{\varepsilon + eA_0}{2mc^2} \approx \frac{1}{2mc^2}\cdot\frac{\mathbf{p}^2}{2m},$$

we eventually obtain

$$\left(\frac{\mathbf{p}^2}{2m} - \frac{\mathbf{p}^4}{8m^3c^2} - eA_0 + \frac{ie\hbar A_0'}{4m^2c^2}\frac{\mathbf{r}\cdot\mathbf{p}}{r} - \frac{eA_0'}{2m^2c^2r}\mathbf{s}\cdot\mathbf{L}\right)\psi_A = \varepsilon\psi_A,$$

in which $\mathbf{s} = \hbar\boldsymbol{\sigma}/2$ is spin and $\mathbf{L} = \mathbf{r}\times\mathbf{p}$ is the orbital angular momentum of the electron. This scenario to proceed to the nonrelativistic limit was outlined by Dirac.

According to an interpretation of the obtained equation, the first two terms follow from a classical expansion

$$\varepsilon = \sqrt{m^2c^4 + c^2\mathbf{p}^2} - mc^2 = \frac{\mathbf{p}^2}{2m} - \frac{\mathbf{p}^4}{8m^3c^2} + \cdots$$

that represents the kinetic energy of the electron. The third term $-eA_0$ is the potential energy of interaction with an external electric field. Quantity

$$\frac{ie\hbar}{4m^2c^2}A'_0\frac{\mathbf{r}\cdot\mathbf{p}}{r} = \frac{e\hbar^2}{4m^2c^2}A'_0\frac{\partial}{\partial r}$$

has no classical analogue. The latter term

$$-\frac{eA'_0}{2m^2c^2 r}\mathbf{s}\cdot\mathbf{L}$$

describes a spin−orbital interaction important for physics; factor $1/2$ appears here in a natural manner, not artificially as the theory of Pauli and Darwin yields. According to a phenomenological consideration, in the nonrelativistic theory one might also introduce this spin−orbital coupling; for agreement with the experiment, one must manually include the so-called Thomas factor $1/2$. After taking this factor into account, the theory of Pauli and Darwin allows one to obtain the correct equation, which is in agreement with the experiment.

Pauli's theory

In the experiment of Stern and Gerlach, atoms of silver in a narrow beam passed through a strong and inhomogeneous magnetic field. Each atom acquired additional energy $W = -\boldsymbol{\mu}\cdot\mathbf{B}$, in which $\boldsymbol{\mu}$ is the magnetic moment of the atom and $\mathbf{B}$ is the magnetic field vector. As a result of the experiment, on a screen, Stern and Gerlach might have obtained some diffuse image corresponding to a mutual orientation $\boldsymbol{\mu}$ and $\mathbf{B}$. However, this result was not observed; instead, the atomic beam became split, such that on the screen only two images symmetrically disposed with respect to the initial beam were discovered. Atomic rays of alkali metals also had two images; for beams containing atoms of vanadium or manganese or iron, the number of images became more than two.

A beam of hydrogen atoms, which are in the S-state, attracts special interest. In this case, the orbital quantum number ℓ of the electron equals zero; consequently, for the electron, the mechanical angular momentum and the magnetic moment, associated with this angular momentum, are completely lacking. As a result of an experiment, the atomic beam again became split into two components under the influence of the magnetic field. This fact bears witness to two possible orientations for the magnetic moment of the electron. Uhlenbeck and Goudsmit supposed *a posteriori* that the electron possesses an intrinsic angular momentum — spin — and the projection of that spin in a selected direction has only two observable values, $\pm\hbar/2$. The corresponding projection of the magnetic moment likewise has only two values.

For an electron, the existence of spin theoretically follows from the relativistic equation of Dirac, but one might consider spin outside special methods of relativistic quantum theory. According to Pauli, spin is an angular momentum, so that it possesses all properties of angular momentum. The eigenvalues of the squared spin angular momentum

$$\mathbf{s}^2 = s_x^2 + s_y^2 + s_z^2$$

are thus $\hbar^2 s(s+1)$; s_x, s_y, and s_z are the projections of spin vector $\mathbf{s}$, and s is the spin quantum number. For each elementary particle, the value of s might be defined only from an experiment, for example, $s = 1/2$ for an electron, proton, neutron, and μ-meson, $s = 0$ for a π-meson, and $s = 1$ for a photon. In the selected representation, one might also determine one projection of spin, for instance, s_z. The possible values for the spin projection number $2s + 1$ in total. For a particle with spin one-half, we have two values; these are eigenvalues of variable s_z that equal $\pm\hbar/2$. The classical limit $\hbar \to 0$ yields zero for spin. Classical mechanics fails to explain the presence of the intrinsic angular momentum for these particles; all models involving a spinning top become absurd and yield nothing useful.

To introduce spin into the nonrelativistic theory, one must consider the wave equation for the electron in an external magnetic field, with a condition that the electron initially has an intrinsic magnetic moment

$$\boldsymbol{\mu} = -\frac{e}{mc}\mathbf{s},$$

in which c is the speed of light, e is the absolute charge of the electron, and m is its mass. We begin from Schrödinger's equation,

$$i\hbar\frac{\partial}{\partial t}|\varphi\rangle = H|\varphi\rangle$$

for states φ; t denotes time. For operator H, we choose the classical expression for a Hamiltonian describing the electron in an external field with vector potential $\mathbf{A}$ and scalar potential U; that is,

$$H = \frac{1}{2m}\left(\mathbf{p} + \frac{e}{c}\mathbf{A}\right)^2 - eU.$$

Adding to this expression the energy of interaction between the electron magnetic moment and the magnetic field, which is characterized by vector $\mathbf{B}$,

$$W = -\boldsymbol{\mu}\cdot\mathbf{B},$$

we obtain

$$H = \frac{1}{2m}\left(\mathbf{p}+\frac{e}{c}\mathbf{A}\right)^2 - eU - \boldsymbol{\mu}\cdot\mathbf{B}.$$

Thus,

$$i\hbar\frac{\partial}{\partial t}|\varphi\rangle = \left[\frac{1}{2m}\left(\mathbf{p}+\frac{e}{c}\mathbf{A}\right)^2 - eU + \frac{e}{mc}\mathbf{s}\cdot\mathbf{B}\right]|\varphi\rangle.$$

Pauli obtained this equation, which describes a motion of the electron in an external electromagnetic field. Pauli's equation is readily generalized for the case of another elementary particle that possesses nonzero spin.

We consider in detail the case $s = 1/2$. With **s**, it is convenient to introduce a new quantity $\boldsymbol{\sigma}(\sigma_x, \sigma_y, \sigma_z)$:

$$\mathbf{s} = \frac{\hbar}{2}\boldsymbol{\sigma};$$

for $\boldsymbol{\sigma}$, we have $\boldsymbol{\sigma}\times\boldsymbol{\sigma} = 2i\boldsymbol{\sigma}$. Because s_z has eigenvalues $\pm\hbar/2$, component σ_z possesses values ± 1, and σ_z^2 has only one value, $+1$. Thus,

$$\sigma_x^2 = \sigma_y^2 = \sigma_z^2 = 1.$$

Using this equality, we find

$$[\sigma_y^2, \sigma_z] = [1, \sigma_z] = 0.$$

Also,

$$[\sigma_y^2, \sigma_z] = \sigma_y[\sigma_y, \sigma_z] + [\sigma_y, \sigma_z]\sigma_y,$$

but $[\sigma_y, \sigma_z] = 2i\sigma_x$, such that

$$\sigma_y\sigma_x + \sigma_x\sigma_y = 0 \text{ or } \sigma_y\sigma_x = -\sigma_x\sigma_y.$$

Hence, σ_x and σ_y commute with an opposite sign — that is, they anticommute. The same conclusions occur for other variables:

$$\sigma_x\sigma_y = -\sigma_y\sigma_x = i\sigma_z,$$
$$\sigma_z\sigma_x = -\sigma_x\sigma_z = i\sigma_y,$$
$$\sigma_y\sigma_z = -\sigma_z\sigma_y = i\sigma_x.$$

To determine an explicit form σ, we recall the formulae obtained previously for the nonzero matrix elements of raising operator L_+ and lowering operator L_- of angular momentum $\mathbf{L}(L_x, L_y, L_z)$:

$$\langle \ell, k \pm 1 | L_\pm | \ell k \rangle = \hbar \sqrt{(\ell \mp k)(\ell \pm k + 1)}.$$

Here, ℓ is a quantum number that characterizes squared angular momentum $\mathbf{L}^2$ and k correspondingly for projection L_z. As $L_x = (L_+ + L_-)/2$ and $L_y = (L_+ - L_-)/2\mathrm{i}$, then

$$\langle \ell, k+1 | L_x | \ell k \rangle = \frac{1}{2} \langle \ell, k+1 | L_+ | \ell k \rangle = \frac{\hbar}{2} \sqrt{(\ell - k)(\ell + k + 1)}$$

and

$$\langle \ell, k+1 | L_y | \ell k \rangle = \frac{1}{2\mathrm{i}} \langle \ell, k+1 | L_+ | \ell k \rangle = -\frac{\mathrm{i}\hbar}{2} \sqrt{(\ell - k)(\ell + k + 1)},$$

in which $\langle \ell, k+1 | L_- | \ell k \rangle = 0$. In an analogous manner,

$$\langle \ell k | L_x | \ell, k+1 \rangle = \frac{1}{2} \langle \ell k | L_- | \ell, k+1 \rangle = \frac{\hbar}{2} \sqrt{(\ell - k)(\ell + k + 1)}$$

and

$$\langle \ell k | L_y | \ell, k+1 \rangle = -\frac{1}{2\mathrm{i}} \langle \ell k | L_- | \ell, k+1 \rangle = \frac{\mathrm{i}\hbar}{2} \sqrt{(\ell - k)(\ell + k + 1)}.$$

We apply these formulae to spin one-half. Assume that $\mathbf{L} = \hbar\boldsymbol{\sigma}/2$, $\ell = s$, and let quantity k retain the preceding meaning of the quantum number of the z-component of angular momentum. We have

$$\langle s, k+1 | \sigma_x | sk \rangle = \langle sk | \sigma_x | s, k+1 \rangle = \sqrt{(s - k)(s + k + 1)}$$

and

$$\langle s, k+1 | \sigma_y | sk \rangle = -\langle sk | \sigma_y | s, k+1 \rangle = -\mathrm{i}\sqrt{(s - k)(s + k + 1)};$$

moreover,

$$\langle sk | s_z | sk \rangle = \hbar k \quad \text{and} \quad \langle sk | \sigma_z | sk \rangle = 2k;$$

$s = 1/2$, whereas $k = \pm 1/2$; one might consequently represent the components of quantity σ in a form of 2×2 Pauli matrices,

$$\sigma_x = \begin{pmatrix} 0 & 1 \\ 1 & 0 \end{pmatrix}, \quad \sigma_y = \begin{pmatrix} 0 & -i \\ i & 0 \end{pmatrix}, \quad \sigma_z = \begin{pmatrix} 1 & 0 \\ 0 & -1 \end{pmatrix},$$

$$\sigma_x^2 = \sigma_y^2 = \sigma_z^2 = \begin{pmatrix} 1 & 0 \\ 0 & 1 \end{pmatrix}.$$

The spin variables separately commute with coordinates x, y, and z, and also with the components of momentum. For a particle with spin half, the commuting variables, for instance, in the coordinate representation, in a complete set therefore become

x, y, z, and σ_z.

Because σ_z has only two values, ± 1, instead of one-component wave function $\langle xyz\sigma_z|\varphi\rangle$, it is convenient to apply a two-component vector,

$$\begin{pmatrix} \langle xyz, +1|\varphi\rangle \\ \langle xyz, -1|\varphi\rangle \end{pmatrix},$$

which is called a spinor. A spinor is hence a function of three, not four, variables.

We proceed to consider the operator of total angular momentum

$$\mathbf{J} = \mathbf{L} + \mathbf{s},$$

$$J_x = L_x + s_x, \quad J_y = L_y + s_y, \quad J_z = L_z + s_z.$$

Because orbital angular momentum $\mathbf{L}$ acts on space coordinates, and $\mathbf{s}$ acts on spin variables, one might satisfy commutative relations

$$[L_r, s_f] = 0, \quad [\mathbf{J}^2, \mathbf{L}^2] = 0, \quad \text{and} \quad [\mathbf{J}^2, \mathbf{s}^2] = 0,$$

in which r and f can equal x or y or z. Quantity $\mathbf{J}$ retains the general properties that exist for angular momentum; hence,

$$\mathbf{J} \times \mathbf{J} = i\hbar \mathbf{J}$$

and

$$[J_x, \mathbf{J}^2] = [J_y, \mathbf{J}^2] = [J_z, \mathbf{J}^2] = 0.$$

The eigenvalues of J_z, by definition, equal $\hbar k_j$, and of $\mathbf{J}^2$ equal $\hbar^2 j(j+1)$. Number j is expressible through orbital and spin quantum numbers ℓ and s:

$$j = |\ell - s|, |\ell - s| + 1, \ldots, \ell + s - 1, \ell + s.$$

For instance, if $s = 1/2$, then $j = 1/2, 3/2, 5/2, \ldots$ and $k_j = \pm 1/2, \pm 3/2, \ldots, \pm j$.

The values for the z-projection of **J** are obtainable directly through the addition of L_z and s_z, such that

$$k_j = k_\ell + k_s,$$

in which quantum number k_ℓ corresponds to the orbital angular momentum with $2\ell + 1$ values and k_s to the spin with $2s + 1$ values. For given values of ℓ and s there must be, in total,

$$(2\ell + 1)(2s + 1)$$

various states. The maximally possible value of k_j equals $\ell + s$; only one state corresponds to this value. Hence, the maximum of j is also equal to $\ell + s$. Decreasing k_j by unity, we obtain $k_j = \ell + s - 1$ and two states

$$\{k_\ell = \ell, \ k_s = s - 1\} \text{ and } \{k_\ell = \ell - 1, \ k_s = s\}$$

that correspond to this value. Number j has two values, $j = \ell + s$ and $j = \ell + s - 1$ at $k_j = \ell + s - 1$. Continuing this scenario with the condition that $s \leq \ell$, we arrive at the value

$$k_j = \ell - s$$

with states of total number $2s + 1$. The minimum of j is thus equal to $\ell - s$. According to a classical point of view, in this case, the vectors **L** and **s** are antiparallel to each other, whereas the maximum value $\ell + s$ corresponds to a parallel orientation of angular momenta **L** and **s**. Notice that if we continued to decrease k_j by unity, we could not obtain new states; as before, their total number at given ℓ and s equals

$$\sum_{j=\ell-s}^{\ell+s} (2j + 1) = (2\ell + 1)(2s + 1).$$

Elementary consequences

In seeking physical solutions in contemporary quantum electrodynamics, the Dirac equation fell from favor. It is preferable to work with the formalism of a scattering matrix and with the powerful methods of perturbation theory, reasoning from only elementary consequences of a one-electron Dirac theory. As one such consequence, a solution of a free Dirac equation primarily persists; this solution in the form of *plane waves* describes the motion of a free particle.

In the absence of an external field, we write the Dirac equation

$$\left(i\hbar\gamma^\mu \frac{\partial}{\partial x^\mu} - mc\right)\psi = 0.$$

We represent a wave function ψ as the product of a four-component spinor u and an exponential factor, that is,

$$\psi = u \cdot \exp(-ip_\nu x^\nu/\hbar),$$

in which p_ν are c-numbers and not yet operators. In this case, the system of differential equations is replaced by the algebraic system

$$(\gamma^\mu p_\mu - mc)u = 0.$$

We multiply this relation by $\gamma^\mu p_\mu + mc$ on the left and take into account that

$$\gamma^\mu \gamma^\nu p_\mu p_\nu = \frac{1}{2}(\gamma^\mu \gamma^\nu + \gamma^\nu \gamma^\mu)p_\mu p_\nu = g^{\mu\nu} p_\mu p_\nu = p_\mu p^\mu;$$

as a result,

$$(p_\mu p^\mu - m^2 c^2)u = 0.$$

Our interest is certainly a nontrivial solution, that is, $u \neq 0$; hence,

$$p_\mu p^\mu - m^2 c^2 = 0$$

or, in an explicit form,

$$\frac{E^2}{c^2} = \mathbf{p}^2 + m^2 c^2,$$

therefore

$$E_\pm = \pm c\sqrt{\mathbf{p}^2 + m^2 c^2}.$$

Together with the states of positive energy E_+, the Dirac equation thus describes the states with negative energy E_-.

We now determine a form for spinor u, assuming

$$u = \begin{pmatrix} u_{1,2} \\ u_{3,4} \end{pmatrix},$$

Physics of the electron

in which

$$u_{1,2} = \begin{pmatrix} u_1 \\ u_2 \end{pmatrix} \text{ and } u_{3,4} = \begin{pmatrix} u_3 \\ u_4 \end{pmatrix}.$$

So,

$$(\gamma^0 p_0 - \gamma \cdot \mathbf{p} - mc)u = 0;$$

on using the explicit form of γ-matrices, we have

$$\left\{ \frac{E}{c} \begin{pmatrix} I & 0 \\ 0 & -I \end{pmatrix} - \sum_{r=1}^{3} p_r \begin{pmatrix} 0 & \sigma_r \\ -\sigma_r & 0 \end{pmatrix} - mc \begin{pmatrix} I & 0 \\ 0 & I \end{pmatrix} \right\} \begin{pmatrix} u_{1,2} \\ u_{3,4} \end{pmatrix} = 0,$$

in which σ_r are 2×2 Pauli matrices. Having readily multiplied the matrices by the spinor, we arrive at the following system of equations:

$$\begin{cases} (E - mc^2)u_1 - c(p_x - ip_y)u_4 - cp_z u_3 = 0, \\ (E - mc^2)u_2 - c(p_x + ip_y)u_3 + cp_z u_4 = 0, \\ -(E + mc^2)u_3 + c(p_x - ip_y)u_2 + cp_z u_1 = 0, \\ -(E + mc^2)u_4 + c(p_x + ip_y)u_1 - cp_z u_2 = 0. \end{cases}$$

We solve this system applying the following simple considerations. In the limiting case, the Dirac equation becomes coincident with the Pauli equation. For a positive value of energy E, that is E_+, we must obtain the *unit* spinors

$$u_{1,2} = \begin{pmatrix} 1 \\ 0 \end{pmatrix} \text{ and } \begin{pmatrix} 0 \\ 1 \end{pmatrix}.$$

So, neglecting the terms of type v/c in our system, in which v is the speed of a particle, we obtain

$$(E - mc^2)u_1 = 0 \text{ and } (E - mc^2)u_2 = 0,$$

hence

$$u_{1,2} = \begin{pmatrix} 1 \\ 0 \end{pmatrix} \text{ and } \begin{pmatrix} 0 \\ 1 \end{pmatrix}.$$

Two other equations yield a negative value of E,

$$(E + mc^2)u_3 = 0 \text{ and } (E + mc^2)u_4 = 0,$$

but in this case, we also treat the unit spinors

$$u_{3,4} = \begin{pmatrix} 1 \\ 0 \end{pmatrix} \text{ and } \begin{pmatrix} 0 \\ 1 \end{pmatrix}.$$

We initially consider the case of positive values of energy for which $E = E_+$. It is convenient to use the first two equations of our system. Using $u_1 = 1$ and $u_2 = 0$, we find

$$\begin{cases} (E_+ - mc^2) - c(p_x - ip_y)u_4 - cp_z u_3 = 0, \\ -c(p_x + ip_y)u_3 + cp_z u_4 = 0, \end{cases}$$

therefore

$$u_3 = \frac{cp_z}{E_+ + mc^2} \text{ and } u_4 = \frac{c(p_x + ip_y)}{E_+ + mc^2},$$

because

$$p_x^2 + p_y^2 + p_z^2 = E_+^2/c^2 - m^2c^2.$$

The first two equations of the system also yield

$$u_1 = 0, \quad u_2 = 1, \quad u_3 = \frac{c(p_x - ip_y)}{E_+ + mc^2}, \quad u_4 = -\frac{cp_z}{E_+ + mc^2}.$$

For $E = E_+$, there is evidently no other linearly independent solution. We proceed to consider the remaining two equations. We assume

$$E = E_-, \quad u_3 = 1 \text{ and } u_4 = 0;$$

then

$$\begin{cases} -(E_- + mc^2) + c(p_x - ip_y)u_2 + cp_z u_1 = 0, \\ c(p_x + ip_y)u_1 - cp_z u_2 = 0. \end{cases}$$

Here, $u_2 = u_1(p_x + ip_y)/p_z$; for u_1, we consequently have the equation

$$-(E_- + mc^2)p_z + c\mathbf{p}^2 u_1 = 0.$$

Taking into account that

$$c^2\mathbf{p}^2 = E_+^2 - m^2c^4 = (E_+ - mc^2)(E_+ + mc^2) = -(E_- + mc^2)(-E_- + mc^2),$$

we find

$$u_1 = \frac{-cp_z}{-E_- + mc^2} \quad \text{and} \quad u_2 = \frac{-c(p_x + ip_y)}{-E_- + mc^2}.$$

The second linearly independent solution is obtainable in an analogous manner; as a result,

$$u_1 = \frac{-c(p_x - ip_y)}{-E_- + mc^2}, \quad u_2 = \frac{cp_z}{-E_- + mc^2}, \quad u_3 = 0, \quad u_4 = 1.$$

One must be aware that the obtained functions are not normalized, which we readily correct. As

$$\sum_{s=1}^{4} |u_s|^2 = 1,$$

and, in our case,

$$\sum_{s=1}^{4} |u_s|^2 = 1 + \frac{c^2 \mathbf{p}^2}{(E_+ + mc^2)^2} = \frac{2E_+}{E_+ + mc^2},$$

then, choosing a normalization factor in a form

$$\sqrt{\frac{E_+ + mc^2}{2E_+}},$$

we find the obtained solution satisfactory.

For the case of a free particle, the Dirac equation thus has an exact solution in a form of plane waves. Two spinors correspond to positive values of energy, and the other two correspond to negative values of energy. Each spinor consists of two small and two large components, and the small components disappear in a nonrelativistic limit. This simplest solution was used in the construction of quantum electrodynamics in the language of a scattering matrix, whether correct or incorrect. In any case, Dirac's theory is a one-electron theory, which is unable to accept all possible processes. In contrast, one commonly uses the "clear" rules of the game, for which there is difficult to find a general scheme resembling a physical theory in its original treatment.

Useful definitions

Until this point we have worked with spinors, which have been represented in the form of columns; the free Dirac equation appeared as

$$(p_0 - \boldsymbol{\alpha} \cdot \mathbf{p} - mc\alpha_m)\psi = 0,$$

but instead of ρ_3, for greater clarity, we introduce the matrix

$$\alpha_m = \rho_3.$$

Spinor ψ is the ket vector. Together with this vector, one must define a complex conjugate spinor — bra vector —

$$\psi^+ = (\psi_1^*, \psi_2^*, \psi_3^*, \psi_4^*);$$

that is, ψ^+ is already a row, not a column. In the Hermitian conjugate Dirac equation, spinor ψ^+ must appear on the left, not the right; that is,

$$\psi^+(-p_0 + \mathbf{p} \cdot \boldsymbol{\alpha} - mc\alpha_m) = 0,$$

because $\alpha_m^+ = \alpha_m$ and $\alpha_s^+ = \alpha_s$, $s = 1, 2$, and 3. Differential operators p_0 and $\mathbf{p}$ act on function ψ^+, which is located on the left, in a standard manner. Having preliminarily multiplied by c, we consequently have, in an explicit form,

$$i\hbar \frac{\partial}{\partial t}\psi + i\hbar c\left(\boldsymbol{\alpha} \cdot \frac{\partial}{\partial \mathbf{r}}\right)\psi - mc^2 \alpha_m \psi = 0$$

and

$$-i\hbar \frac{\partial}{\partial t}\psi^+ - i\hbar c\left(\frac{\partial}{\partial \mathbf{r}}\psi^+ \cdot \boldsymbol{\alpha}\right) - mc^2 \psi^+ \alpha_m = 0.$$

We multiply the former equation by ψ^+ on the left, the latter by ψ on the right, and subtract one from the other to obtain

$$\frac{\partial}{\partial t}(\psi^+ \psi) + \mathrm{div}(c\psi^+ \boldsymbol{\alpha}\psi) = 0.$$

Introducing the definitions for a probability density

$$\rho = \psi^+ \psi$$

and a probability current

$$\mathbf{j} = c\psi^+ \boldsymbol{\alpha}\psi,$$

we rewrite the obtained equality in the conventional form of a continuity equation

$$\frac{\partial \rho}{\partial t} + \mathrm{div}\,\mathbf{j} = 0.$$

We convert this equation to a covariant form, replacing α with $\gamma^0\gamma$,

$$\frac{\partial j^\mu}{\partial x^\mu} = 0,$$

in which j^μ is the four-vector of probability current that is equal to

$$j^\mu = c\psi^+ \gamma^0 \gamma^\mu \psi \equiv c\overline{\psi}\gamma^\mu\psi;$$

$\overline{\psi} = \psi^+ \gamma^0$ is the Dirac adjoint spinor.

Positrons

Proceeding to an important physical consequence, we consider the solutions of the Dirac equation in an external field,

$$\left(\left(p_0 + \frac{e}{c}A_0\right) - \sum_{s=1}^{3}\alpha_s\left(p_s + \frac{e}{c}A_s\right) - mc\alpha_m\right)\psi = 0,$$

corresponding to negative values of energy. It is convenient to choose a representation for α-matrices such that α_1, α_2, and α_3 are real and α_m is complex. In preceding definitions, one should simply interchange α_2 and α_m. We assume that this operation has been effected in our equation. We invoke a complex conjugation of the equation, using $i \to -i$. Then, $p_\mu \to -p_\mu$, $\alpha_s \to \alpha_s$, $\alpha_m \to -\alpha_m$, $\psi \to \psi^*$, and all components A_μ of the four-vector of a potential of a field remain unaltered; as a result,

$$\left(\left(p_0 - \frac{e}{c}A_0\right) - \sum_{s=1}^{3}\alpha_s\left(p_s - \frac{e}{c}A_s\right) - mc\alpha_m\right)\psi^* = 0.$$

When ψ corresponds to negative energy, ψ^* corresponds to positive values of energy. Replacing e with $-e$ in the second equation, one might obtain a formal coincidence of both equations. The Dirac equation thus equally describes the electrons and new particles of electronic mass and charge $+e$. The entire question is whether these states — those of a new particle — represent electronic states belonging to a negative spectrum of energy. Why are the electrons unable to spontaneously fall into states with a negative energy? Dirac supposed that all or nearly all states with a negative energy are occupied. Hence, the vacuum is the state with entirely occupied levels of a negative energy and unoccupied levels of a positive energy. The appearance of an additional electron involves only an occupation of a state of positive energy, because the transition to the negative spectrum of energy is forbidden through Pauli's principle. Furthermore, an unoccupied state with a negative energy implies that there exists a new particle of positive energy and charge

+e. According to Dirac, the unoccupied states with a negative energy incarnate the new particles—positrons. Such an explanation is sometimes called a theory of *holes*. A hole is the absence of an electron with a negative energy.

The process of the birth of a positron and an electron, for instance, might be interpreted as a transition of an electron that absorbs radiation from a state with a negative value of energy to a state with a positive value of energy. However, in the case of annihilation, an electron jumps to the unoccupied state with a negative energy; hence, the electron−positron couple disappears with the appearance of radiation. All this action is comprehensively confirmed in experiments. That a positron was predicted before the factual discovery of the new particle heralded a total triumph for Dirac's theory.

Fine structure

For a hydrogen atom, the Dirac equation has an exact solution, like the Schrödinger equation. We consider in detail a case of a centrally symmetric field in Dirac's theory of the hydrogen atom.

We write Dirac's Hamiltonian

$$H = -eA_0(r) + c\rho_1(\boldsymbol{\sigma} \cdot \mathbf{p}) + mc^2\rho_3,$$

which describes the motion of an electron in a field with a Coulombic potential

$$A_0(r) = \frac{e}{r};$$

hence, r becomes the distance between an electron and a proton in a hydrogen atom. Let E and ψ be the eigenvalues and eigenfunctions of Hamiltonian H, then

$$\left(-E - \frac{e^2}{r} + c\rho_1(\boldsymbol{\sigma} \cdot \mathbf{p}) + mc^2\rho_3\right)\psi = 0.$$

We multiply this equation on the left by

$$E + \frac{e^2}{r} + c\rho_1(\boldsymbol{\sigma} \cdot \mathbf{p}) + mc^2\rho_3$$

and take into account that

$$\rho_3\rho_1 + \rho_1\rho_3 = 0$$

and

$$[1/r, \boldsymbol{\sigma} \cdot \mathbf{p}] = \boldsymbol{\sigma} \cdot i\hbar\nabla(1/r) = -\frac{i\hbar}{r^3}\boldsymbol{\sigma} \cdot \mathbf{r};$$

as a result,

$$\left(-\left(E+\frac{e^2}{r}\right)^2 - \frac{i\hbar c e^2}{r^3}\rho_1(\boldsymbol{\sigma}\cdot\mathbf{r}) + c^2\mathbf{p}^2 + m^2c^4\right)\psi = 0.$$

In this equation, we proceed to spherical coordinates but involve a proviso that in a central force field the total angular momentum

$$\mathbf{J} = \mathbf{L} + \frac{\hbar}{2}\boldsymbol{\sigma}$$

is preserved, not the orbital angular momentum $\mathbf{L}$ as in Schrödinger's theory. For this purpose, we apply the method of noncommutative algebra proposed by Dirac.[2]

So, as

$$\mathbf{p}^2 = -\frac{\hbar^2}{r^2}\frac{\partial}{\partial r}\left(r^2\frac{\partial}{\partial r}\right) + \frac{\mathbf{L}^2}{r^2},$$

then assuming

$$\psi = \frac{v}{r},$$

we transform our equation to the form

$$\left(-\left(E+\frac{e^2}{r}\right)^2 - \frac{i\hbar c e^2}{r^3}\rho_1(\boldsymbol{\sigma}\cdot\mathbf{r}) + c^2\left(p_r^2 + \frac{\mathbf{L}^2}{r^2}\right) + m^2c^4\right)v = 0,$$

in which $p_r = -i\hbar\partial/\partial r$. Furthermore, on using the formula

$$(\boldsymbol{\sigma}\cdot\mathbf{a})(\boldsymbol{\sigma}\cdot\mathbf{b}) = \mathbf{a}\cdot\mathbf{b} + i\boldsymbol{\sigma}\cdot(\mathbf{a}\times\mathbf{b}),$$

which is valid for arbitrary vectors $\mathbf{a}$ and $\mathbf{b}$, we calculate $(\boldsymbol{\sigma}\cdot\mathbf{L})^2$:

$$(\boldsymbol{\sigma}\cdot\mathbf{L})^2 = \mathbf{L}^2 + i\boldsymbol{\sigma}\cdot(\mathbf{L}\times\mathbf{L}) = \mathbf{L}^2 - \hbar\boldsymbol{\sigma}\cdot\mathbf{L} = \left(\mathbf{L}+\frac{\hbar}{2}\boldsymbol{\sigma}\right)^2 - 2\hbar\boldsymbol{\sigma}\cdot\mathbf{L} - \frac{3}{4}\hbar^2,$$

therefore

$$(\boldsymbol{\sigma}\cdot\mathbf{L}+\hbar)^2 = \mathbf{J}^2 + \frac{\hbar^2}{4}.$$

Moreover,

$$(\boldsymbol{\sigma}\cdot\mathbf{L})(\boldsymbol{\sigma}\cdot\mathbf{p}) = i\boldsymbol{\sigma}\cdot(\mathbf{L}\times\mathbf{p}) \text{ and } (\boldsymbol{\sigma}\cdot\mathbf{p})(\boldsymbol{\sigma}\cdot\mathbf{L}) = i\boldsymbol{\sigma}\cdot(\mathbf{p}\times\mathbf{L}),$$

and, on summing these equalities, one obtains

$$i\sigma \cdot ((\mathbf{L} \times \mathbf{p}) + (\mathbf{p} \times \mathbf{L})) = i\sum_{1,2,3} \sigma_1([L_2,p_3] + [p_2,L_3]) = 2i^2\hbar \sum_{1,2,3} \sigma_1 p_1 = -2\hbar(\sigma \cdot \mathbf{p});$$

consequently,

$$(\sigma \cdot \mathbf{L} + \hbar)(\sigma \cdot \mathbf{p}) + (\sigma \cdot \mathbf{p})(\sigma \cdot \mathbf{L} + \hbar) = 0.$$

Quantity $\sigma \cdot \mathbf{L} + \hbar$ commutes with all terms of Dirac's Hamiltonian except term $c\rho_1(\sigma \cdot \mathbf{p})$, which anticommutes. Because $\rho_3\rho_1 + \rho_1\rho_3 = 0$, assuming therefore $j\hbar = \rho_3(\sigma \cdot \mathbf{L} + \hbar)$, we obtain a new quantity j that commutes with H and thus represents a constant of motion. In addition, as $\rho_3^2 = 1$, square $j^2\hbar^2$ equals $\mathbf{J}^2 + \hbar^2/4$ such that, according to the formulae for the addition of angular momenta, the eigenvalues of j are integers of both negative and positive values, eliminating zero.

In an analogous manner, we find that

$$[\rho_1(\sigma \cdot \mathbf{r}), j] = 0,$$

because $\mathbf{r}$ and $\mathbf{p}$ enter with a proper symmetry into moment $\mathbf{L}$. To simplify quantity $\rho_1(\sigma \cdot \mathbf{r})$, we introduce a matrix λ:

$$\lambda r = \rho_1(\sigma \cdot \mathbf{r}).$$

Obviously, r commutes with both ρ_1 and $\sigma \cdot \mathbf{r}$; hence,

$$\lambda^2 r^2 = (\lambda r)^2 = \rho_1^2(\sigma \cdot \mathbf{r})^2 = r^2,$$

therefore

$$\lambda^2 = 1.$$

Let us consider the commutator

$$[\rho_1(\sigma \cdot \mathbf{r}), \mathbf{r} \cdot \mathbf{p}].$$

On the one side,

$$[\rho_1(\sigma \cdot \mathbf{r}), \mathbf{r} \cdot \mathbf{p}] = \rho_1\sigma \cdot [\mathbf{r}, \mathbf{r} \cdot \mathbf{p}] = i\hbar\rho_1(\sigma \cdot \mathbf{r});$$

on the other side,

$$[\rho_1(\sigma \cdot \mathbf{r}), \mathbf{r} \cdot \mathbf{p}] = [\lambda r, r p_r] = i\hbar\lambda r + r^2[\lambda, p_r],$$

in which $rp_r = \mathbf{r} \cdot \mathbf{p}$; on comparison, we find that

$$[\lambda, p_r] = 0.$$

Consequently, quantity λ commutes with both r and p_r.
Now with

$$\mathbf{L}^2 = (\boldsymbol{\sigma} \cdot \mathbf{L})(\boldsymbol{\sigma} \cdot \mathbf{L} + \hbar) = \hbar^2 \rho_3 j(\rho_3 j - 1)$$

and

$$\rho_1(\boldsymbol{\sigma} \cdot \mathbf{r}) = \lambda r,$$

we return to the equation and we have

$$\left(-E^2 - \frac{2Ee^2}{r} + \frac{c^2\hbar^2 \rho_3 j(\rho_3 j - 1) - i\hbar c e^2 \lambda - e^4}{r^2} + c^2 p_r^2 + m^2 c^4\right) v = 0.$$

Introducing the quantity

$$\alpha = \frac{e^2}{\hbar c},$$

we rewrite this equation in the form

$$\left(p_r^2 - \frac{2Ee^2}{c^2 r} + \frac{\hbar^2}{r^2}(j^2 - \alpha^2 - \rho_3 j - i\alpha\lambda)\right) v = \left(\frac{E^2}{c^2} - m^2 c^2\right) v.$$

Matrices ρ_3 and λ commute with all other variables in the equation and anticommute with each other. Assuming that

$$\Gamma = \rho_3 j + i\alpha\lambda,$$

we obtain

$$\left(p_r^2 - \frac{2Ee^2}{c^2 r} + \frac{\hbar^2 \Gamma(\Gamma - 1)}{r^2}\right) v = \left(\frac{E^2}{c^2} - m^2 c^2\right) v,$$

in which introduced quantity Γ commutes with all variables;

$$\Gamma^2 = j^2 - \alpha^2.$$

Operator j, being a constant of motion, can be replaced by its eigenvalues

$$j = \pm 1, \pm 2, \ldots$$

Hence,

$$\Gamma_\pm = \pm\sqrt{j^2 - \alpha^2}$$

are the eigenvalues of operator Γ; Γ_+ corresponds to positive values of j, whereas Γ_- corresponds to negative values of j. Proceeding to the eigenvalues for Γ, we eventually obtain

$$\left(p_r^2 - \frac{2Ee^2}{c^2 r} + \frac{\hbar^2 \Gamma_\pm (\Gamma_\pm - 1)}{r^2}\right) v = \left(\frac{E^2}{c^2} - m^2 c^2\right) v.$$

This equation, essentially representing a relativistic radial equation, determines the possible energy levels of Dirac's hydrogen atom.

Solution according to factorization

To solve the radial equation, we apply the method of factorization, for which purpose the eigenvalues of the operator must be defined

$$F = p_r^2 + \frac{\hbar^2 \Gamma_\pm (\Gamma_\pm - 1)}{r^2} - \frac{2\zeta}{r}, \quad \zeta = \frac{Ee^2}{c^2}.$$

Supposing a_n and b_n to be real c-numbers, we use

$$\eta_n = p_r + i\left(a_n + \frac{b_n}{r}\right).$$

It is easily seen that

$$\eta_n^+ \eta_n = (p_r - i(a_n + b_n/r))(p_r + i(a_n + b_n/r)) = p_r^2 + a_n^2 + \frac{2 a_n b_n}{r} + \frac{b_n^2 - \hbar b_n}{r^2}$$

and

$$\eta_n \eta_n^+ = p_r^2 + a_n^2 + \frac{2 a_n b_n}{r} + \frac{b_n^2 + \hbar b_n}{r^2}.$$

For F_1, we therefore have

$$F_1 = \eta_1^+ \eta_1 + f_1 = p_r^2 + a_1^2 + \frac{2 a_1 b_1}{r} + \frac{b_1^2 - \hbar b_1}{r^2} + f_1.$$

As $F_1 = F$, then

$$a_1 b_1 = -\zeta, \quad b_1(b_1 - \hbar) = \hbar^2 \Gamma_\pm (\Gamma_\pm - 1) \quad \text{and} \quad a_1^2 + f_1 = 0.$$

Two variants for a solution are appropriate. If

$$b_1 = \hbar \Gamma_\pm,$$

then

$$a_1 = -\frac{\zeta}{\hbar \Gamma_\pm} \quad \text{and} \quad f_1 = -\frac{\zeta^2}{\hbar^2 \Gamma_\pm^2}.$$

Otherwise

$$b_1 = \hbar(1 - \Gamma_\pm),$$

and then

$$a_1 = -\frac{\zeta}{\hbar(1 - \Gamma_\pm)} \quad \text{and} \quad f_1 = -\frac{\zeta^2}{\hbar^2(1 - \Gamma_\pm)^2}.$$

For Γ_+, we choose $b_1 = \hbar \Gamma_+$, because in this case

$$-\frac{\zeta^2}{\hbar^2(1-\Gamma_+)^2} < -\frac{\zeta^2}{\hbar^2 \Gamma_+^2};$$

for Γ_-, obviously, $b_1 = \hbar(1 - \Gamma_-)$.

From a comparison

$$\eta_n \eta_n^+ + f_n = \eta_{n+1}^+ \eta_{n+1} + f_{n+1}$$

of two expressions for F_{n+1}, we find

$$a_n b_n = a_{n+1} b_{n+1},$$

$$b_n(b_n + \hbar) = b_{n+1}(b_{n+1} - \hbar),$$

and

$$a_n^2 + f_n = a_{n+1}^2 + f_{n+1}.$$

Solution $b_n = -b_{n+1}$ has no physical meaning, because in this case

$$a_n = -a_{n+1} \quad \text{and} \quad f_n = f_{n+1}.$$

If
$$b_{n+1} = b_n + \hbar = b_{n-1} + 2\hbar = \cdots = b_1 + n\hbar,$$

then
$$a_n = -\frac{\zeta}{b_n} = -\frac{\zeta}{\hbar(n-1+\Gamma_+)} \quad \text{for } j > 0$$

and
$$a_n = -\frac{\zeta}{b_n} = -\frac{\zeta}{\hbar(n-\Gamma_-)} \quad \text{for } j < 0.$$

Furthermore,
$$a_{n+1}^2 + f_{n+1} = a_n^2 + f_n = \cdots = a_1^2 + f_1 = 0;$$

therefore, $f_n = -a_n^2$. On the other side, the eigenvalues of operator F are expressible as
$$f_n = \frac{E^2}{c^2} - m^2 c^2.$$

Consequently,
$$-\frac{\alpha^2 E^2}{c^2(s+\sqrt{j^2-\alpha^2})^2} = \frac{E^2}{c^2} - m^2 c^2,$$

in which $s = n - 1 = 0, 1, 2, \ldots$ for positive values of j or $s = n = 1, 2, 3, \ldots$ for negative values of j. Expressing E, we obtain
$$E = mc^2 \left(1 + \frac{\alpha^2}{(s+\sqrt{j^2-\alpha^2})^2}\right)^{-1/2}.$$

For the levels of a hydrogen atom in a relativistic case, there exists an exact solution in which arises a *fine structure constant*
$$\alpha = \frac{e^2}{\hbar c} \approx \frac{1}{137},$$

which is sufficiently small; therefore, with good accuracy, $\sqrt{j^2 - \alpha^2} \approx |j|$ and
$$E = mc^2 - \frac{me^4}{2\hbar^2} \cdot \frac{1}{(s+|j|)^2} + \cdots$$

Subtracting mc^2 from here, we see that in first order, with respect to α^2, the obtained relativistic expression becomes coincident with Bohr's nonrelativistic formula.

Unfortunately, this exact solution is not free of an accidental degeneracy. In states with the same number j but different orbital quantum numbers ℓ, one and the same value of energy E pertains. Further, the interaction of an electron with its "own" field of radiation eliminates this degeneracy. Such a subtle electromagnetic effect, which is called a Lamb shift, yields a splitting of otherwise coincident levels. For instance, the developed theory of the Dirac's hydrogen atom predicts the degeneracy of states

$2S_{1/2}$ and $2P_{1/2}$,

whereas according to the experiment of Lamb and Retherford, level $2S_{1/2}$ lies slightly above level $2P_{1/2}$.

Magnetic interaction

The Dirac equation supports exact solutions for a freely moving electron and for the case of the Coulomb potential, but other exact solutions exist. For instance, one might readily obtain energy levels for an electron in a homogeneous magnetic field. According to the nonrelativistic Schrödinger theory, we have a similar structure of energy levels, which are known as Landau levels.

Understanding the specifics of the forthcoming solution, at least by analogy with Landau levels, we consider an electron in a constant magnetic field as a particular case of a general electronic motion, which is described with the equation

$$\left(\gamma^\mu \left(p_\mu + \frac{e}{c}A_\mu\right) - mc\right)\psi = 0,$$

in which

$$p_\mu = (p_0, -\mathbf{p}) = \left(\frac{E}{c}, i\hbar\frac{\partial}{\partial \mathbf{r}}\right)$$

is four-momentum,

$$A_\mu = (0, 0, -A_y(x), 0)$$

is four-potential of the electromagnetic field, E is the energy of the electron, e is the absolute charge of the electron, and m is its mass. We suppose that the magnetic field vector is directed along axis z, and its magnitude has a weak dependence on x, almost constant. In a sense, the Dirac electron in a weakly inhomogeneous magnetic field is a curious anharmonicity effect.

To begin our calculations, we choose Dirac matrices; let these be

$$\gamma^0 = \begin{pmatrix} I & 0 \\ 0 & -I \end{pmatrix}, \gamma_r = \begin{pmatrix} 0 & \sigma_r \\ -\sigma_r & 0 \end{pmatrix},$$

in which σ_r are 2×2 Pauli matrices,

$$\sigma_x = \begin{pmatrix} 0 & 1 \\ 1 & 0 \end{pmatrix}, \ \sigma_y = \begin{pmatrix} 0 & -i \\ i & 0 \end{pmatrix}, \ \sigma_z = \begin{pmatrix} 1 & 0 \\ 0 & -1 \end{pmatrix};$$

and I is a 2×2 unit matrix. Let us write the equation

$$\left\{ \gamma^0(E/c) - \gamma_x p_x - \gamma_y \left(p_y + \frac{e}{c} A_y \right) - \gamma_z p_z - mc \right\} \begin{pmatrix} \psi_{1,2} \\ \psi_{3,4} \end{pmatrix} = 0.$$

For brevity, we introduce here a typical notation for a spinor as

$$\psi_{1,2} = \begin{pmatrix} \psi_1 \\ \psi_2 \end{pmatrix} \text{ and } \psi_{3,4} = \begin{pmatrix} \psi_3 \\ \psi_4 \end{pmatrix}.$$

Taking into account that

$$\gamma^0 \psi = \begin{pmatrix} \psi_{1,2} \\ -\psi_{3,4} \end{pmatrix}, \ \gamma_r \psi = \begin{pmatrix} \sigma_r \psi_{3,4} \\ -\sigma_r \psi_{1,2} \end{pmatrix},$$

and further that

$$\sigma_x \psi_{1,2} = \begin{pmatrix} \psi_2 \\ \psi_1 \end{pmatrix}, \ \sigma_y \psi_{1,2} = i \begin{pmatrix} -\psi_2 \\ \psi_1 \end{pmatrix}, \ \sigma_z \psi_{1,2} = \begin{pmatrix} \psi_1 \\ -\psi_2 \end{pmatrix},$$

in an analogous manner

$$\sigma_x \psi_{3,4} = \begin{pmatrix} \psi_4 \\ \psi_3 \end{pmatrix}, \ \sigma_y \psi_{3,4} = i \begin{pmatrix} -\psi_4 \\ \psi_3 \end{pmatrix}, \ \sigma_z \psi_{3,4} = \begin{pmatrix} \psi_3 \\ -\psi_4 \end{pmatrix},$$

and we obtain the system of equations for the spinor components:

$$\frac{E}{c} \psi_1 - p_x \psi_4 + i \left(p_y + \frac{e}{c} A_y \right) \psi_4 - p_z \psi_3 - mc \psi_1 = 0,$$

$$\frac{E}{c} \psi_2 - p_x \psi_3 - i \left(p_y + \frac{e}{c} A_y \right) \psi_3 + p_z \psi_4 - mc \psi_2 = 0,$$

$$-\frac{E}{c} \psi_3 + p_x \psi_2 - i \left(p_y + \frac{e}{c} A_y \right) \psi_2 + p_z \psi_1 - mc \psi_3 = 0,$$

$$-\frac{E}{c}\psi_4 + p_x\psi_1 + i\left(p_y + \frac{e}{c}A_y\right)\psi_1 - p_z\psi_2 - mc\psi_4 = 0.$$

Through a condition that potential A_y depends on x only, the solution becomes chosen in the form

$$\psi_j = u_j(x) \cdot \exp\left(\frac{ip_y y}{\hbar} + \frac{ip_z z}{\hbar}\right).$$

As a result, we have

$$f^- u_1 + i\hbar\frac{\partial u_4}{\partial x} + iPu_4 - p_z u_3 = 0,$$

$$f^- u_2 + i\hbar\frac{\partial u_3}{\partial x} - iPu_3 + p_z u_4 = 0,$$

$$-f^+ u_3 - i\hbar\frac{\partial u_2}{\partial x} - iPu_2 + p_z u_1 = 0,$$

$$-f^+ u_4 - i\hbar\frac{\partial u_1}{\partial x} + iPu_1 - p_z u_2 = 0,$$

in which $f^\pm = (E/c) \pm mc$, $P = p_y + (e/c)A_y$, and p_y and p_z are c-numbers. We express u_3 and u_4 from the latter two equations,

$$u_3 = \frac{1}{f^+}\left(-i\hbar\frac{\partial u_2}{\partial x} - iPu_2 + p_z u_1\right), \quad u_4 = \frac{1}{f^+}\left(-i\hbar\frac{\partial u_1}{\partial x} + iPu_1 - p_z u_2\right),$$

and substitute them into the former two equations; as a result,

$$\frac{\partial^2 u_1}{\partial x^2} + \frac{1}{\hbar^2}\left(f^- f^+ - p_z^2 - P^2 - \hbar\frac{\partial P}{\partial x}\right)u_1 = 0,$$

$$\frac{\partial^2 u_2}{\partial x^2} + \frac{1}{\hbar^2}\left(f^- f^+ - p_z^2 - P^2 + \hbar\frac{\partial P}{\partial x}\right)u_2 = 0.$$

Equations for u_3 and u_4 yield the same result:

$$\frac{\partial^2 u_3}{\partial x^2} + \frac{1}{\hbar^2}\left(f^- f^+ - p_z^2 - P^2 - \hbar\frac{\partial P}{\partial x}\right)u_3 = 0,$$

$$\frac{\partial^2 u_4}{\partial x^2} + \frac{1}{\hbar^2}\left(f^- f^+ - p_z^2 - P^2 + \hbar\frac{\partial P}{\partial x}\right)u_4 = 0.$$

These equations correspond to negative values of energy and two possible projections of the electron spin.

Omitting the index of function u_j, we eventually combine equations to obtain

$$\frac{\partial^2 u}{\partial x^2} + \frac{1}{\hbar^2}\left(\frac{E^2}{c^2} - m^2 c^2 - p_z^2 - \left(p_y + \frac{e}{c}A_y\right)^2 - \hbar e \frac{\sigma}{c}\frac{\partial A_y}{\partial x}\right) u = 0,$$

in which $\sigma = +1$ for components u_1 and u_3 and $\sigma = -1$ for components u_2 and u_4.

Landau levels

We apply a condition that the magnitude of the field is almost constant and expand A_y in powers of x:

$$A_y = a_1 x + a_2 x^2 + \cdots = \sum_{i>0} a_i x^i.$$

Coefficients a_i for $i > 1$ are assumed to be sufficiently small that in zero-order approximation, the z-projection of the magnetic field vector equals $\partial A_y / \partial x = a_1 = \text{const}$. Moreover, we introduce, by definition, a function

$$F(x) = \left(p_y + \frac{e}{c}A_y\right)^2 + \hbar e \frac{\sigma}{c}\frac{\partial A_y}{\partial x}$$

and transform $F(x)$:

$$p_y + \frac{e}{c}A_y = \sum_i b_i x^i, \quad b_0 = p_y, \quad b_{i>0} = e\frac{a_i}{c};$$

$$\left(p_y + \frac{e}{c}A_y\right)^2 = \left(\sum_i b_i x^i\right)^2 = \sum_k B_k x^k, \quad B_k = b_0 b_k + b_1 b_{k-1} + b_2 b_{k-2} + \cdots + b_k b_0;$$

$$\hbar e \frac{\sigma}{c}\frac{\partial A_y}{\partial x} = \hbar e \frac{\sigma}{c}\sum_k (k+1) a_{k+1} x^k.$$

Therefore, if $C_k = B_k + \hbar e (\sigma/c)(k+1) a_{k+1}$, then $F(x) = \sum_k C_k x^k$.

Let small coefficients a_i with $i > 1$ be coefficients of an anharmonic type such that there is a linear transformation $q = x - x_0$; then

$$F(x) = \sum_k C_k x^k = \sum_k Q_k q^k = F(q),$$

with $\partial F / \partial x|_{x=x_0} = 0$, i.e., $Q_1 = 0$; the arbitrary coefficients have the form

$$Q_k = \frac{1}{k!}F^{(k)}(q)_{q=0} = \frac{1}{k!}F^{(k)}(x)_{x=x_0} = \frac{1}{k!}\sum_j C_j j(j-1)\ldots(j-k+1)x_0^{j-k}.$$

We see that $Q_1 = \sum_j C_j j x_0^{j-1} = 0$; from this equation follows the value of x_0; then, we find $Q_0 = \sum_j C_j x_0^j$ and other quantities Q_k.

We thus obtain equation

$$\frac{\partial^2 u}{\partial q^2} + \left(\kappa - \frac{Q_2}{\hbar^2}q^2 - \frac{Q_3}{\hbar^2}q^3 - \frac{Q_4}{\hbar^2}q^4 - \cdots\right)u = 0,$$

$$\kappa = \frac{1}{\hbar^2}\left(\frac{E^2}{c^2} - m^2c^2 - p_z^2 - Q_0\right).$$

One should primarily investigate a particular solution $a_{i>1} = 0$; in this case,

$$b_0 = p_y, \quad b_1 = e\frac{a_1}{c}, \quad B_0 = p_y^2, \quad B_1 = 2e\frac{a_1}{c}p_y, \quad B_2 = \left(\frac{ea_1}{c}\right)^2,$$

$$C_0 = B_0 + \hbar e\frac{\sigma}{c}a_1, \quad C_1 = B_1, \quad C_2 = B_2.$$

Furthermore,

$$Q_1 = C_1 + 2C_2 x_0 = 0;$$

therefore,

$$x_0 = -\frac{cp_y}{ea_1}$$

and

$$Q_0 = C_0 + C_1 x_0 + C_2 x_0^2 = \hbar e\frac{\sigma}{c}a_1, \quad Q_2 = C_2 = \left(\frac{ea_1}{c}\right)^2;$$

other coefficients are equal to zero. Assuming $q = \sqrt{c\hbar/ea_1}\,\xi$, we obtain the equation of a harmonic oscillator,

$$\frac{\partial^2 u}{\partial \xi^2} + \left(\frac{c\hbar}{ea_1}\kappa - \xi^2\right)u = 0,$$

which indicates that $(c\hbar/ea_1)\kappa = 2n + 1$; hence,

$$E^2 = m^2c^4 + c^2p_z^2 + c\hbar ea_1(2n+1+\sigma), \quad n = 0, 1, 2, \ldots,$$

in which $\sigma = +1$ for components u_1 and u_3 and $\sigma = -1$ for components u_2 and u_4; a_1 is the strength of the magnetic field. All functions u_1, u_2, u_3, and u_4 become expressible through Hermite polynomials.

The exact solution of the Dirac equation for an electron in a homogeneous magnetic field is thus derived. The expression for energy includes two classical terms, $(mc^2)^2$ and $(cp_z)^2$, plus a quantized quantity, which appears through the motion in the plane perpendicular to axis z. The electron momentum directed along the magnetic field vector retains continuous values, and the rotatory motion in plane xy is described with energy levels of a harmonic oscillator type.

Anharmonicity

To return to our general problem, we take into account the anharmonicity. Introducing new variable ξ through the relation $q = (\hbar/\sqrt{Q_2})^{1/2}\xi \equiv \lambda\xi$, we have

$$-\frac{\partial^2 u}{\partial \xi^2} + \xi^2 u + \sum_{k>2}\frac{Q_k}{\hbar^2}\lambda^{k+2}\xi^k u = (\lambda^2 \kappa)u.$$

Assuming

$$H^0 = -\frac{\partial^2}{2\partial\xi^2} + \frac{1}{2}\xi^2, \quad c_p = \frac{Q_{p+2}}{2\hbar^2}\lambda^4 = \frac{Q_{p+2}}{2Q_2}, \quad \varepsilon = \frac{1}{2}(\lambda^2\kappa) = \frac{\hbar}{2\sqrt{Q_2}}\kappa,$$

obviously

$$\left(H^0 + \sum_{p>0} c_p \lambda^p \xi^{p+2}\right) u = \varepsilon u.$$

This equation is typical in the theory of anharmonicity; as $Q_2 \gg Q_3, Q_4, \ldots$, the solution might be expressible through a series of perturbation theory, for instance, in a form of polynomials of quantum numbers. The zero-order approximation is a simple harmonic oscillator. In this case, $c_{p>0} = 0$, $\varepsilon_0 = n + 1/2$, $n = 0, 1, 2, \ldots$, and functions $u(\xi)$ are expressed in terms of Hermite polynomials. From the general structure of an anharmonic Hamiltonian, arbitrary corrections to ε_0 clearly depend only on quantum number n in a polynomial manner. Each correction ε_α of order α ($\alpha = 1, 2, \ldots$) is therefore a sum of some polynomials, which depend on coefficients c_p parametrically. These corrections add to unperturbed quantity ε_0, which represents the square of the energy; as a result, we obtain some expansion in n or $n + 1/2$, i.e.,

$$\varepsilon = \sum_i \zeta_i (n+1/2)^i.$$

Coefficients ζ_i are defined from a general solution with perturbation theory, which is given by a simple expression[5]

$$\varepsilon = \varepsilon_0 + \sum_\alpha \lambda^\alpha \varepsilon_\alpha, \quad \varepsilon_\alpha = \frac{1}{\alpha} \sum_{(p\beta\gamma)\alpha} pc_p \Pi_{\beta\gamma}^{p+2}(n,n), \quad p=1,2,\ldots, \quad \beta,\gamma=0,1,2,\ldots,$$

in which $(p\beta\gamma)\alpha$ denotes a summation over indices p, β, and γ under a condition in which $p+\beta+\gamma=\alpha$. Quantities $\Pi_{\beta\gamma}^s(n,n)$ are polynomials of quantum numbers, which follows from the *recurrence* relations and can be taken from the table (see Chapter 3). For instance, the first-order correction equals zero; the second-order correction is

$$\varepsilon_2 = -(30c_1^2 - 6c_2)(n+1/2)^2 + \text{const, etc.}$$

In conclusion, we note that one might consider the Klein−Fock−Gordon equation in a similar manner. For a particle with zero spin and charge e' in a weak inhomogeneous magnetic field, this equation has the form

$$\left(\frac{E^2}{c^2} - m^2c^2 - p_x^2 - \left(p_y - \frac{1}{c}e'A_y\right)^2 - p_z^2\right)\psi = 0.$$

With $p_x = -i\hbar\partial/\partial x$, $p_y = p_y$, $p_z = p_z$, we obtain

$$\frac{\partial^2 \varphi}{\partial x^2} + \frac{1}{\hbar^2}\left(\frac{E^2}{c^2} - m^2c^2 - p_z^2 - P^2\right)\varphi = 0, \quad \psi = \varphi(x)\exp\left(\frac{ip_y y}{\hbar} + \frac{ip_z z}{\hbar}\right).$$

This equation is a particular case of the equation in the Dirac theory considered here; simply, $\sigma = 0$. Other calculations remain valid.

Theory of anharmonicity

Model Hamiltonian

We proceed to study the general questions of the theory of anharmonicity with a discussion of a model anharmonic oscillator, a Hamiltonian of which has the form

$$H = H^0 + \hbar\omega a_1 \lambda \xi^3 + \hbar\omega a_2 \lambda^2 \xi^4,$$

in which H^0 is the Hamiltonian of a harmonic oscillator that makes small vibrations with frequency ω; parameter λ reflects the order of smallness of anharmonic coefficients

a_1 and a_2;

ξ is the vibrational variable that in terms of the operators for creation η^+ and destruction η is expressible as

$$\xi = \eta + \eta^+.$$

Eigenvalues E_n^0 and eigenvectors $|n\rangle$ of operator H^0 are known already, namely,

$$E_n^0 = \hbar\omega\left(n + \frac{1}{2}\right), \quad n = 0, 1, 2, \ldots$$

We know also the relations

$$\eta^+|n-1\rangle = \sqrt{n}|n\rangle \text{ and } \eta|n\rangle = \sqrt{n}|n-1\rangle,$$

which yield the result of action on state $|n\rangle$ with the operators of creation and destruction, respectively. With their aid, one might calculate the matrix elements of vibrational variable to various first powers, for instance,

$$\langle m|\xi|n\rangle = \sqrt{n} \cdot \delta_{m,n-1} + \sqrt{n+1} \cdot \delta_{m,n+1},$$
$$\langle m|\xi^2|n\rangle = \sqrt{n(n-1)} \cdot \delta_{m,n-2} + (2n+1) \cdot \delta_{m,n} + \sqrt{(n+1)(n+2)} \cdot \delta_{m,n+2},$$
$$\langle m|\xi^3|n\rangle = \sqrt{n(n-1)(n-2)} \cdot \delta_{m,n-3} + 3n^{3/2} \cdot \delta_{m,n-1} + 3(n+1)^{3/2} \cdot \delta_{m,n+1}$$
$$+ \sqrt{(n+1)(n+2)(n+3)} \cdot \delta_{m,n+3}$$

and

$$\langle m|\xi^4|n\rangle = \sqrt{n(n-1)(n-2)(n-3)} \cdot \delta_{m,n-4} + 2(2n-1)\sqrt{n(n-1)} \cdot \delta_{m,n-2}$$
$$+ 3(2n^2 + 2n + 1) \cdot \delta_{m,n} + 2(2n+3)\sqrt{(n+1)(n+2)} \cdot \delta_{m,n+2}$$
$$+ \sqrt{(n+1)(n+2)(n+3)(n+4)} \cdot \delta_{m,n+4}.$$

Regarding eigenvalues E_n and eigenvectors $|\psi_n\rangle$ of Hamiltonian H, they can be found only approximately. We consider the calculation of E_n and $|\psi_n\rangle$ in a framework of perturbation theory.

According to the general results, which are shown in Chapter 1, the sought eigenvalues and eigenvectors are determined from the system of equations

$$\frac{d}{d\lambda} E_n(\lambda) = \langle \psi_n(\lambda)|W|\psi_n(\lambda)\rangle,$$
$$\frac{d}{d\lambda}|\psi_n(\lambda)\rangle = \sum_{m \neq n} \frac{\langle \psi_m(\lambda)|W|\psi_n(\lambda)\rangle}{E_n(\lambda) - E_m(\lambda)} |\psi_m(\lambda)\rangle;$$

here,

$$W = \frac{\partial H'}{\partial \lambda} = \hbar\omega a_1 \xi^3 + 2\hbar\omega a_2 \lambda \xi^4,$$

because in our case

$$H' = \hbar\omega a_1 \lambda \xi^3 + \hbar\omega a_2 \lambda^2 \xi^4.$$

Assuming

$$E_n(\lambda) = E_n^0 + \sum_{\alpha=1}^{\infty} \lambda^\alpha E_n^\alpha \quad \text{and} \quad |\psi_n(\lambda)\rangle = |n\rangle + \sum_{\alpha=1}^{\infty} \lambda^\alpha |n, \alpha\rangle,$$

in which E_n^α and $|n, \alpha\rangle$ are the corresponding corrections, we rewrite the equations of the system in the form

$$E_n^1 + 2\lambda E_n^2 + \cdots = \hbar\omega \langle \psi_n(\lambda)|(a_1\xi^3 + 2a_2\lambda\xi^4)|\psi_n(\lambda)\rangle,$$
$$|n, 1\rangle + 2\lambda|n, 2\rangle + \cdots = \hbar\omega \sum_{m \neq n} \frac{\langle \psi_m(\lambda)|(a_1\xi^3 + 2a_2\lambda\xi^4)|\psi_n(\lambda)\rangle}{E_n(\lambda) - E_m(\lambda)} |\psi_m(\lambda)\rangle,$$

hence, at $\lambda = 0$, the first-order corrections follow immediately, namely,

$$E_n^1 = \hbar\omega a_1 \langle n|\xi^3|n\rangle = 0$$

and

$$|n,1\rangle = \hbar\omega a_1 \sum_{m\neq n} \frac{\langle m|\xi^3|n\rangle}{E_n^0 - E_m^0}|m\rangle = a_1\left(\frac{1}{3}\sqrt{n(n-1)(n-2)}|n-3\rangle + 3n^{3/2}|n-1\rangle\right.$$
$$\left. - 3(n+1)^{3/2}|n+1\rangle - \frac{1}{3}\sqrt{(n+1)(n+2)(n+3)}|n+3\rangle\right).$$

To obtain the corrections of the second order, one must differentiate the equations of the system with respect to λ and then set λ equal to zero; as a result,

$$E_n^2 = \frac{1}{2}\hbar\omega a_1(\langle n,1|\xi^3|n\rangle + \langle n|\xi^3|n,1\rangle) + \hbar\omega a_2\langle n|\xi^4|n\rangle$$

and

$$|n,2\rangle = \frac{1}{2}\hbar\omega a_1 \sum_{m\neq n} \frac{\langle m,1|\xi^3|n\rangle + \langle m|\xi^3|n,1\rangle}{E_n^0 - E_m^0}|m\rangle + \frac{1}{2}\hbar\omega a_1 \sum_{m\neq n} \frac{\langle m|\xi^3|n\rangle}{E_n^0 - E_m^0}|m,1\rangle$$
$$+ \hbar\omega a_2 \sum_{m\neq n} \frac{\langle m|\xi^4|n\rangle}{E_n^0 - E_m^0}|m\rangle.$$

For E_n^2, obviously

$$E_n^2 = -\hbar\omega a_1^2(30n^2 + 30n + 11) + \hbar\omega a_2(6n^2 + 6n + 3).$$

Regarding $|n,2\rangle$, we only note that this correction represents the expansion with respect to vectors $|n-k\rangle$ and $|n+k\rangle$, in which $k = 0, 2, 4$, and 6.

The calculation of corrections of higher order, beginning with $\alpha = 3$, has no meaning because for the correct calculation of these corrections, one must add to the model Hamiltonian the terms containing

$$\xi^5, \xi^6, \ldots$$

Thus, one might conclude that the principal problem of the theory of anharmonicity consists of the investigation of the Hamiltonian

$$H = H^0 + \hbar\omega \sum_{p>0} \lambda^p a_p \xi^{p+2},$$

in which a_p are anharmonic coefficients. We proceed to solve this problem.

Perturbation method

Let us solve a general problem for the eigenvalues and eigenvectors of anharmonic Hamiltonian H. If there is only one variable, then, as we already know,

$$H = H^0 + \hbar\omega \sum_{p>0} \lambda^p a_p \xi^{p+2};$$

in the case of r normal vibrations (see section 'Quantum numbers' in Chapter 1),

$$H = H^0 + \sum_{p>0} \lambda^p \sum_{(j_1 j_2 \ldots j_r) p+2} a_{j_1 j_2 \ldots j_r} \xi_1^{j_1} \xi_2^{j_2} \ldots \xi_r^{j_r}$$

or

$$H = H^0 + \sum_{p>0} G_p(\xi) \lambda^p,$$

in which ξ implies all vibrational variables $\xi_1, \xi_2, \ldots, \xi_r$. Operator H^0, representing a harmonic Hamiltonian, is characterized by eigenvalues E_n^0 and eigenvectors $|n\rangle$.

We summarize the basic requirements for the computational formalism of the anharmonicity problem. First, the concurrence of separate orders must be taken into account correctly, and the contribution of each perturbation group G_p to the sought result must be considered. The first perturbation order is determined by the quantity G_1, the second is determined by G_1 and G_2, and so on. Second, the recurrent character of perturbation theory must be taken advantage of, and algebraic expressions for corrections of higher order must be derived from the lowest approximations. This approach allows one to avoid repeated calculations, because information regarding the perturbation is already involved in the preceding approximation to which there is no need to return. Third, difficulties of renormalization of the wavefunction when proceeding from a current correction to that of the next order must be overcome. This requirement provides more subtle work with experimental data. Finally, to save all expressions in a clear manner, the final formulae must not be bulky.

We differentiate the equation for eigenvalues $E_n(\lambda)$ and eigenfunctions $|n, \lambda\rangle$ of Hamiltonian H with respect to parameter λ:

$$\sum_p p G_p(\xi) \lambda^{p-1} |n, \lambda\rangle - \frac{\partial E_n(\lambda)}{\partial \lambda} |n, \lambda\rangle = (E_n(\lambda) - H) \frac{\partial}{\partial \lambda} |n, \lambda\rangle.$$

Vector $|n, \lambda\rangle$, which is terminated with a parenthesis, characterizes the exact state. For an infinitesimal change in parameter λ, we obtain vector $|n, \lambda + \delta\lambda\rangle$, which is represented as a series expansion $\sum_m A_{mn}(\delta\lambda)|m, \lambda\rangle$; that is,

$$|n, \lambda + \delta\lambda\rangle = A_{nn}(\delta\lambda)|n, \lambda\rangle + \sum_{m \neq n} A_{mn}(\delta\lambda)|m, \lambda\rangle.$$

This expansion is universally valid by virtue of the completeness of the eigenfunctions. Coefficients A_{mn} are related through an expression for normalization

$$|A_{nn}|^2 + \sum_{m \neq n} |A_{mn}|^2 = 1.$$

When $\delta\lambda = 0$, we have $A_{nn} = 1$ and $A_{mn} = 0$; hence, there exists a nonzero limit for the ratio $A_{mn}/\delta\lambda$ as $\delta\lambda \to 0$, which is, by definition, equal to C_{mn}. As a result, we obtain

$$A_{nn} = 1 - \text{const} \cdot \delta\lambda^2 + \cdots$$

and

$$\frac{\partial}{\partial\lambda}|n, \lambda\rangle = \lim_{\delta\lambda \to 0} \frac{|n, \lambda + \delta\lambda\rangle - |n, \lambda\rangle}{\delta\lambda} = \sum_{m \neq n} C_{mn}|m, \lambda\rangle.$$

Returning to the differentiated equation for the eigenvalues and eigenfunctions, we have

$$\sum_p p G_p(\xi)\lambda^{p-1}|n, \lambda\rangle - \frac{\partial E_n(\lambda)}{\partial\lambda}|n, \lambda\rangle = \sum_{m \neq n} C_{mn}(E_n(\lambda) - E_m(\lambda))|m, \lambda\rangle.$$

Using this equation, we find

$$C_{mn} = \sum_p p\lambda^{p-1} \frac{\langle m, \lambda|G_p(\xi)|n, \lambda\rangle}{E_n(\lambda) - E_m(\lambda)}$$

and determine the exact expansions for $\partial|n, \lambda\rangle/\partial\lambda$ and $\partial E_n(\lambda)/\partial\lambda$ with respect to parameter λ:

$$\frac{\partial E_n(\lambda)}{\partial\lambda} = \sum_p p\lambda^{p-1}\langle n, \lambda|G_p(\xi)|n, \lambda\rangle,$$

$$\frac{\partial}{\partial\lambda}|n, \lambda\rangle = \sum_{m \neq n}\sum_p p\lambda^{p-1} \frac{\langle m, \lambda|G_p(\xi)|n, \lambda\rangle}{E_n(\lambda) - E_m(\lambda)}|m, \lambda\rangle.$$

Introducing the sought corrections E_n^α and $|n, \alpha\rangle$ of perturbation theory through these series expansions

$$E_n(\lambda) = E_n^0 + \sum_{\alpha > 0}\lambda^\alpha E_n^\alpha \text{ and } |n, \lambda\rangle = |n\rangle + \sum_{\alpha > 0}\lambda^\alpha|n, \alpha\rangle,$$

we obtain

$$\sum_{\alpha>0} \alpha \lambda^{\alpha-1} E_n^\alpha = \sum_{p\beta\gamma} p\lambda^{p+\beta+\gamma-1} \langle n,\beta|G_p(\xi)|n,\gamma\rangle,$$

$$\sum_{\alpha>0} \alpha \lambda^{\alpha-1} |n,\alpha\rangle = \sum_{pq\beta\gamma\nu}\sum_{m\neq n} p\lambda^{p+q+\beta+\gamma+\nu-1} \Delta_q(n,m)\langle m,\beta|G_p(\xi)|n,\gamma\rangle|m,\nu\rangle,$$

therefore[6]

$$E_n^\alpha = \frac{1}{\alpha}\sum_{(p\beta\gamma)\alpha} p\langle n,\beta|G_p(\xi)|n,\gamma\rangle,$$

$$|n,\alpha\rangle = \frac{1}{\alpha}\sum_{(pq\beta\gamma\nu)\alpha}\sum_{m\neq n} p\Delta_q(n,m)\langle m,\beta|G_p(\xi)|n,\gamma\rangle|m,\nu\rangle,$$ (3.1)

in which

$$\Delta_q(n,m) = \frac{1}{q!}\frac{\partial^q}{\partial \lambda^q}\left[\frac{1}{E_n(\lambda)-E_m(\lambda)}\right]_{\lambda=0}$$

$$= \frac{1}{E_n^0-E_m^0}\sum_i\sum_{(\alpha_1\alpha_2\ldots\alpha_i)q}\prod_{e=1}^i \frac{E_m^{\alpha_e}-E_n^{\alpha_e}}{E_n^0-E_m^0},\quad \Delta_0(n,m) = \frac{1}{E_n^0-E_m^0}.$$

Here, $\alpha>0$ and the relation for $\Delta_q(n,m)$ is based on a simple expansion $(1-x)^{-1} = 1+x+x^2+\cdots$, which holds for $|x|<1$.

For a one-dimensional case, we have $G_p = \hbar\omega\, a_p \xi^{p+2}$; according to Eq. (3.1), the first-order corrections to the vibrational energy and the corresponding function are determined by the matrix elements of quantity $\hbar\omega\, a_1\xi^3$. Clearly, $E_n^1 = 0$ and

$$|n,1\rangle = a_1\left(\frac{1}{3}\sqrt{n(n-1)(n-2)}|n-3\rangle + 3n^{3/2}|n-1\rangle - 3(n+1)^{3/2}|n+1\rangle \right.$$

$$\left. - \frac{1}{3}\sqrt{(n+1)(n+2)(n+3)}|n+3\rangle\right),$$ (3.2)

in which $|n\rangle$ is the state vector of a harmonic oscillator.

For the r-dimensional case, the result in the first order is defined by a more complicated perturbation function

$$\hbar\sum_i \omega_i a_i \xi_i^3 + \hbar\sum_{ij} a_{ij}\xi_i^2\xi_j + \hbar\sum_{ijk} a_{ijk}\xi_i\xi_j\xi_k,$$

in which ω_i denote the harmonic frequencies; a_i, a_{ij}, and a_{ijk} are force parameters in which $a_{ii} = 0$ and $a_{ijk} = 0$ for $i \geq j \geq k$. The first correction to the energy is equal to the diagonal matrix element of the perturbing function. One sees directly that the latter is an odd function; consequently, $E_n^1 = 0$. The first correction $|n, 1\rangle$ to the wavefunction is determined by the nonzero nondiagonal matrix elements of the perturbation. Clearly, $E_n^1 = 0$ and

$$|n, 1\rangle = \sum_i |n_1\rangle\ldots|n_i, 1\rangle\ldots|n_r\rangle + |n, A\rangle + |n, B\rangle + |n, C\rangle, \quad (3.3)$$

in which n implies all quantum numbers from n_1 to n_r. Here, $|n_i, 1\rangle$ is determined by quantity $\hbar\omega_i a_i \xi_i^3$ and is exactly equal to Eq. (3.2). Corrections $|n, A\rangle$ and $|n, B\rangle$ appear through $\hbar\sum_{ij} a_{ij}\xi_i^2 \xi_j$; as a result,

$$|n, A\rangle = \sum_{ij} \frac{2a_{ij}}{\omega_j}\left(n_i + \frac{1}{2}\right)(\eta_j - \eta_j^+)|n\rangle,$$

$$|n, B\rangle = \sum_{ij} a_{ij}\left(\frac{\eta_i^2 \eta_j - (\eta_i^+)^2 \eta_j^+}{2\omega_i + \omega_j} + \frac{\eta_i^2 \eta_j^+ - (\eta_i^+)^2 \eta_j}{2\omega_i - \omega_j}\right)|n\rangle,$$

in which appear operators η_i^+ for creation and η_i for destruction:

$$\eta_i^+|n\rangle = \sqrt{n_i + 1}|n_1, n_2, \ldots, n_i + 1, \ldots, n_r\rangle$$

and

$$\eta_i|n\rangle = \sqrt{n_i}|n_1, n_2, \ldots, n_i - 1, \ldots, n_r\rangle.$$

Taking into account $\hbar \sum_{ijk} a_{ijk}\xi_i\xi_j\xi_k$, we find

$$|n, C\rangle = \sum_{ijk} a_{ijk}\left(\frac{\eta_i\eta_j\eta_k - \eta_i^+\eta_j^+\eta_k^+}{\omega_i + \omega_j + \omega_k} + \frac{\eta_i\eta_j\eta_k^+ - \eta_i^+\eta_j^+\eta_k}{\omega_i + \omega_j - \omega_k} + \frac{\eta_i\eta_j^+\eta_k - \eta_i^+\eta_j\eta_k^+}{\omega_i - \omega_j + \omega_k}\right.$$

$$\left.+ \frac{\eta_i^+\eta_j\eta_k - \eta_i\eta_j^+\eta_k^+}{-\omega_i + \omega_j + \omega_k}\right)|n\rangle.$$

Such a construction of the first correction to the wavefunction becomes convenient and might readily have an interpretation; namely, for normal vibration i, $|n_i, 1\rangle$ characterizes its own anharmonicity. Quantity $|n, A\rangle$ represents an expansion in terms of the states, with each of which containing only a single normal vibration that becomes perturbed. Vectors $|n, B\rangle$ and $|n, C\rangle$ are expansions in states with two and three perturbed vibrations, respectively.

Inclusion of degenerate levels

With regard to degenerate states in the polynomial formalism, we consider whether the developed perturbation theory retains its advantages over the conventional theory. We begin with the general case[7] for which the exact equation for eigenvalues $E_{nx}(\lambda)$ and eigenfunctions $|nx, \lambda\rangle$ has the form

$$\left(H^0 + \sum_{p>0} \lambda^p G_p \right) |nx, \lambda\rangle = E_{nx}(\lambda)|nx, \lambda\rangle.$$

Here, G_p are perturbations of various orders in small parameter λ, and the zero-order Hamiltonian H^0 has degenerate eigenvalues $E_n^0 (= E_{nx}^0)$ with functions $|nx\rangle$, in which index x numbers the degenerate states corresponding to level n. Repeating, to some extent, the reasoning used in our derivation of the basic relations of the perturbation theory, we construct the algebraic solutions for $E_{nx}(\lambda)$ and $|nx, \lambda\rangle$ in the form of series in λ to various powers with the inclusion of degeneracy.

The method is formally simple. Differentiating the exact eigenvalue equation with respect to λ and using an expansion in the exact eigenvectors,

$$\frac{\partial}{\partial \lambda}|nx, \lambda\rangle = \sum_{my \neq nx} C_{my,nx} |my, \lambda\rangle,$$

we generate the following system:

$$\frac{\partial E_{nx}(\lambda)}{\partial \lambda} = \sum_p p\lambda^{p-1} \langle nx, \lambda | G_p | nx, \lambda\rangle,$$

$$\frac{\partial}{\partial \lambda}|nx, \lambda\rangle = \sum_{my \neq nx} \sum_p p\lambda^{p-1} \frac{\langle my, \lambda | G_p | nx, \lambda\rangle}{E_{nx}(\lambda) - E_{my}(\lambda)} |my, \lambda\rangle.$$

Assuming, as customarily, that

$$E_{nx}(\lambda) = E_n^0 + \sum_{\alpha > 0} \lambda^\alpha E_{nx}^\alpha \quad \text{and} \quad |nx, \lambda\rangle = |nx\rangle + \sum_{\alpha > 0} \lambda^\alpha |nx, \alpha\rangle,$$

we obtain the desired expressions for corrections E_{nx}^α and $|nx, \alpha\rangle$ for the degenerate case in a recurrent manner,

$$E_{nx}^\alpha = \frac{1}{\alpha} \sum_{(p\beta\gamma)\alpha} p \langle nx, \beta | G_p | nx, \gamma\rangle,$$

$$|nx, \alpha\rangle = \frac{1}{\alpha} \sum_{y; m \neq n} \sum_{(pq\beta\gamma\nu)\alpha} p\Delta_q^0(nx, my) \langle my, \beta | G_p | nx, \gamma\rangle |my, \nu\rangle \qquad (3.4)$$
$$+ \frac{1}{\alpha} \sum_{y \neq x} \sum_{(pq\beta\gamma\nu)\alpha > \sigma} p\Delta_q^\sigma(nx, ny) \langle ny, \beta | G_p | nx, \gamma\rangle |ny, \nu\rangle,$$

in which $\Delta_0^\sigma(nx, my) = (E_{nx}^\sigma - E_{my}^\sigma)^{-1}$ and, for $q > 0$,

$$\Delta_q^\sigma(nx, my) = \frac{1}{E_{nx}^\sigma - E_{my}^\sigma} \sum_i \sum_{(\alpha_1 \alpha_2 \ldots \alpha_i)q} \prod_{e=1}^i \frac{E_{my}^{\sigma+\alpha_e} - E_{nx}^{\sigma+\alpha_e}}{E_{nx}^\sigma - E_{my}^\sigma}.$$

Factor $\Delta_q^0(nx, my)$ retains its preceding meaning,

$$\frac{1}{E_{nx}(\lambda) - E_{my}(\lambda)} = \frac{1}{(E_n^0 - E_m^0) + (E_{nx}^1 - E_{my}^1)\lambda + \cdots} = \sum_q \lambda^q \Delta_q^0(nx, my),$$

whereas for the degenerate states of level n,

$$\frac{1}{E_{nx}(\lambda) - E_{ny}(\lambda)} = \frac{1}{(E_{nx}^\sigma - E_{ny}^\sigma)\lambda^\sigma + (E_{nx}^{\sigma+1} - E_{ny}^{\sigma+1})\lambda^{\sigma+1} + \cdots} = \sum_q \lambda^{q-\sigma} \Delta_q^\sigma(nx, ny),$$

in which $\sigma \geq 1$. If, for instance, $E_{nx}^1 - E_{ny}^1 \neq 0$ for all $y \neq x$, then $\sigma = 1$. If $E_{nx}^1 - E_{ny}^1 = 0$, then one must consider $\sigma = 2$, etc.

Our theory thus remains valid in the presence of degenerate states. Because the expansion coefficients of $\partial |nx, \lambda\rangle / \partial \lambda$ in terms of $|my, \lambda\rangle$ contain only differences $E_{nx}(\lambda) - E_{my}(\lambda)$ as the denominators, the developed method possesses a further merit: it allows one to eliminate zeros in the denominators on summing over the degenerate states. Comparison of the results obtained here with the nondegenerate case shows a formal similarity of the expressions for E_{nx}^α and an evident distinction in calculating the corrections $|nx, \alpha\rangle$ involving an additional summation over the group of degenerate states beginning with $\alpha = \sigma + 1$.

To elucidate the meaning of σ, we consider the first two corrections in detail; for the first correction,

$$|nx, 1\rangle = \sum_{y; m \neq n} \frac{\langle my|G_1|nx\rangle}{E_n^0 - E_m^0} |my\rangle + \left(\sum_{y \neq x} \frac{\langle ny|G_1|nx\rangle}{E_{nx}^0 - E_{ny}^0} |ny\rangle = 0 \right).$$

The absence of terms with $\sigma = 0$ implies that all matrix elements $\langle ny|G_1|nx\rangle$ vanish for $y \neq x$, which allows one to eliminate all zeros in denominators in the group of degenerate states. This result indicates that the correct functions of zero approximation $|nx\rangle$ were chosen as the basis functions. This operation is effected through an appropriate unitary transformation of the eigenfunctions on the basis of the solution of a secular equation. Assuming that $\sigma = 1$, we consider the second correction

$$|nx, 2\rangle = \frac{1}{2} \sum_{y; m \neq n} \sum_{(pq\beta\gamma\nu)2} p\Delta_q^0(nx, my) \langle my, \beta|G_p|nx, \gamma\rangle |my, \nu\rangle$$

$$+ \frac{1}{2} \sum_{y \neq x} \frac{\langle ny, 1|G_1|nx\rangle + \langle ny|G_1|nx, 1\rangle + 2\langle ny|G_2|nx\rangle}{E_{nx}^1 - E_{ny}^1} |ny\rangle.$$

If for some reason

$$E_{nx}^1 - E_{ny}^1 = \langle nx|G_1|nx \rangle - \langle ny|G_1|ny \rangle = 0,$$

then functions $|nx\rangle$ should again be unitarily transformed (with the aid of matrix U)

$$|nx'\rangle = \sum_x U_{xx'} |nx\rangle$$

so that

$$\langle ny', 1|G_1|nx' \rangle + \langle ny'|G_1|nx', 1 \rangle + 2\langle ny'|G_2|nx' \rangle = 0.$$

We emphasize that, in the new basis set,

$$\langle ny'|G_1|nx' \rangle = \sum_{xy} U_{yy'}^* U_{xx'} \langle ny|G_1|nx \rangle = \sum_x U_{xy'}^* U_{xx'} \langle nx|G_1|nx \rangle = \langle nx|G_1|nx \rangle \delta_{y'x'},$$

i.e., all matrix elements $\langle ny'|G_1|nx' \rangle$ vanish as before, except the case $y' = x'$. Consequently, for $E_{nx}^1 - E_{ny}^1 = 0$, $\sigma = 2$. The summation within the group of degenerate states should be performed only in calculating the corrections $|nx, \alpha\rangle$ of third order and greater; therefore, with $\alpha \geq 3$.

To eliminate trivial zeros in the denominators when $E_{nx}^0 - E_{ny}^0 = 0$, one must thus perform a unitary transformation of the basis set so that, for $y \neq x$,

$$\sum_{(p\beta\gamma)1} p\langle ny, \beta|G_p|nx, \gamma \rangle = \langle ny|G_1|nx \rangle = 0.$$

This basis set was used initially. If the degeneracy is removed, then $\sigma = 1$; otherwise, when $E_{nx}^1 - E_{ny}^1 = 0$, the basis functions should again be transformed so that, in the new basis set,

$$\sum_{(p\beta\gamma)2} p\langle ny', \beta|G_p|nx', \gamma \rangle = 0;$$

consequently, $\sigma = 2$. This reasoning becomes generalized with a simple scheme $(\sigma = \sigma)$:

$$E_{nx}^{\sigma-1} - E_{ny}^{\sigma-1} = 0 \rightarrow \sum_{(p\beta\gamma)\sigma} p\langle ny^{(\sigma-1)}, \beta|G_p|nx^{(\sigma-1)}, \gamma \rangle = 0.$$

After the value of σ is chosen on sequentially eliminating the zeros, the required corrections $|nx, \alpha\rangle$ and E_{nx}^α thus become reconstructed from the equations in Eq. (3.4).

Polynomial formalism

We specify the parity of the harmonic state vector so that $|n \pm k\rangle$ at a fixed quantum number n has the parity of number k; for instance, $|n-1\rangle$ and $|n+3\rangle$ characterize odd states, $|n-2\rangle$, and $|n\rangle$ even states. Moreover, for $m \geq n$, we introduce factor g_{nm} as

$$g_{nm} = (n+1)(n+2)\ldots(m-1)m, \quad g_{nn} = 1.$$

The system of equations (Eq. (3.1)) for the case of one variable, when $\alpha > 0$ and quantity $\Delta_q(n,m)$ is chosen to be dimensionless, takes the following form:

$$E_n^\alpha = \frac{\hbar\omega}{\alpha} \sum_{(p\beta\gamma)\alpha} pa_p \langle n, \beta | \xi^{p+2} | n, \gamma \rangle,$$

$$|n, \alpha\rangle = \frac{1}{\alpha} \sum_{(pq\beta\gamma\nu)\alpha} \sum_{m \neq n} pa_p \Delta_q(n,m) \langle m, \beta | \xi^{p+2} | n, \gamma \rangle | m, \nu \rangle, \qquad (3.5)$$

$$\Delta_q(n,m) = \hbar\omega \sum_i \sum_{(\alpha_1\ldots\alpha_i)q} \frac{(E_m^{\alpha_1} - E_n^{\alpha_1})\ldots(E_m^{\alpha_i} - E_n^{\alpha_i})}{(E_n^0 - E_m^0)^{i+1}}, \quad \Delta_0(n,m) = \frac{\hbar\omega}{E_n^0 - E_m^0}.$$

In particular, it follows that $E_n^1 = 0$ and $|n, 1\rangle$ contains states of only odd parity (see Eq. (3.2)). Neglecting a small constant, for the second-order correction we have

$$E_n^2 = -\hbar\omega(30a_1^2 - 6a_2)\left(n + \frac{1}{2}\right)^2;$$

states of only even parity are involved in function $|n, 2\rangle$.

In the general case, vector $|n, \alpha\rangle$ involves states $|n \pm k\rangle$ of parity α with $k = 3\alpha$ as a bound of the expansion. The necessary condition for this generalization is identity $\Delta_{2q-1} = 0$, which follows from trivial equality $E_n^{2\alpha-1} = 0$. To prove this assertion, it suffices to consider Eq. (3.5) more thoroughly. Under an assumption that function $|n, \gamma\rangle$ is expanded in terms of states $|n \pm c_\gamma\rangle$, in which number c_γ has parity γ, we readily obtain condition $m \pm c_\beta = n \pm c_\gamma \pm c_p$, which must be satisfied by the nonzero matrix elements $\langle m, \beta | \xi^{p+2} | n, \gamma \rangle$. Function $|n, \alpha\rangle$ is therefore an expansion in terms of vectors $|n \pm c_p \pm c_\beta \pm c_\gamma \pm c_\nu\rangle$, whereas function $|n, \alpha\rangle$ is formed by states $|n \pm c_\alpha\rangle$. We consequently have the equality $\pm c_\alpha = \pm c_p \pm c_\beta \pm c_\gamma \pm c_\nu$. This result does not contradict the relation $\alpha = p + q + \beta + \gamma + \nu$ for only even values of q. Quantity Δ_{2q-1} is thus identically equal to zero. The converse statement is also obviously true: if correction $|n, \alpha\rangle$ contains states $|n \pm c_\alpha\rangle$, then we have $E_n^{2\alpha-1} = 0$ for an arbitrary odd correction. This conclusion establishes the validity of this assertion.

To demonstrate this analysis, we rewrite $|n, \alpha\rangle$ in accordance with Eq. (3.5) in the form

$$|n, \alpha\rangle = a_1^\alpha \sum_{m_1 m_2 \ldots m_\alpha} h(n, m, \alpha) \langle n|\xi^3|m_1\rangle \langle m_1|\xi^3|m_2\rangle \ldots \langle m_{\alpha-1}|\xi^3|m_\alpha\rangle |m_\alpha\rangle + \cdots,$$

in which an explicit form of quantity $h(n, m, \alpha)$ has no special interest. We see that vectors $|n \pm 3\alpha\rangle$ are really bound states in an expansion of $|n, \alpha\rangle$ in harmonic state vectors. Moreover, in calculating the amplitude of a harmonic state, for example, state $|n + k\rangle$ of function $|n, \alpha\rangle$, it is necessary to sum various products

$$\langle n|\xi^s|\ell\rangle \langle \ell|\xi^q|r\rangle \ldots \langle u|\xi^p|n + k\rangle,$$

which are proportional to $\sqrt{g_{n,n+k}}$ through a relation $\langle \ell|\xi^q|\ell + k\rangle \sim \sqrt{g_{\ell,\ell+k}}$.

To introduce polynomials $\Pi^s_{\alpha\beta}(n, m)$ of quantum numbers n and m at $m \geq n$ is thus convenient in the following form:

$$\langle n, \alpha|\xi^s|m, \beta\rangle = \sqrt{g_{nm}} \Pi^s_{\alpha\beta}(n, m). \tag{3.6}$$

In this definition, one must distinguish between the orders of indices α and β, and between the orders of numbers n and m.

To convert from matrix elements to polynomials, it suffices to multiply the left side of Eq. (3.5) for $|n, \alpha\rangle$ by expression $\langle \ell, \mu|\xi^s$; i.e.,

$$\langle \ell, \mu|\xi^s|n, \alpha\rangle = \frac{1}{\alpha} \sum_{(pq\beta\gamma\nu)\alpha} \sum_{m \neq n} p a_p \Delta_q(n, m) \langle m, \beta|\xi^{p+2}|n, \gamma\rangle \langle \ell, \mu|\xi^s|m, \nu\rangle.$$

On substituting Eq. (3.6) and making elementary algebraic transformations, we obtain the *recurrence* equations,

$$\Pi^s_{\mu\alpha}(\ell, n) = \frac{1}{\alpha} \sum_{(pq\beta\gamma\nu)\alpha} p a_p \left[\sum_{m < \ell} g_{m\ell} \Delta_q \Pi^s_{\nu\mu}(m, \ell) \Pi^{p+2}_{\beta\gamma}(m, n) \right.$$
$$\left. + \sum_{\ell \leq m < n} \Delta_q \Pi^s_{\mu\nu}(\ell, m) \Pi^{p+2}_{\beta\gamma}(m, n) + \sum_{m > n} g_{nm} \Delta_q \Pi^s_{\mu\nu}(\ell, m) \Pi^{p+2}_{\gamma\beta}(n, m) \right],$$

$$\Pi^s_{\alpha\mu}(n, \ell) = \frac{1}{\alpha} \sum_{(pq\beta\gamma\nu)\alpha} p a_p \left[\sum_{m < n} g_{mn} \Delta_q \Pi^s_{\nu\mu}(m, \ell) \Pi^{p+2}_{\beta\gamma}(m, n) \right.$$
$$\left. + \sum_{n < m \leq \ell} \Delta_q \Pi^s_{\nu\mu}(m, \ell) \Pi^{p+2}_{\gamma\beta}(n, m) + \sum_{m > \ell} g_{\ell m} \Delta_q \Pi^s_{\mu\nu}(\ell, m) \Pi^{p+2}_{\gamma\beta}(n, m) \right].$$

To determine the polynomials, we set $\ell = n - k$ in the first equation and $\ell = n + k$ in the second equation; in this case, the polynomials clearly become expansions in n to various powers.

According to Eq. (3.6), some polynomials equal zero; namely, function $|n \pm k, \beta\rangle$ is expanded in terms of states $|n \pm k \pm c_\beta\rangle$, in which number c_β has parity β. Correspondingly, quantity $\xi^s|n \pm k, \beta\rangle$ is an expansion in terms of harmonic vectors $|n \pm k \pm c_\beta \pm c_s\rangle$. Matrix elements $\langle n, \alpha|\xi^s|n \pm k, \beta\rangle$ become defined by quantities $\langle n \pm c_\alpha|n \pm k \pm c_\beta \pm c_s\rangle$, yielding $k = \pm c_s \pm c_\alpha \pm c_\beta$. In polynomials $\Pi^s_{\beta\alpha}(n - k, n)$ and $\Pi^s_{\alpha\beta}(n, n + k)$, number k therefore has the parity of number $s + \alpha + \beta$. The maximum or bounding value of k is also determined by numbers s, α, and β and equals $3(\alpha + \beta) + s$. Polynomials $\Pi^s_{\beta\alpha}(n - k, n)$ and $\Pi^s_{\alpha\beta}(n, n + k)$ for $k > 3(\alpha + \beta) + s$ hence are identically equal to zero.

Apart from a direct calculation of the polynomials with the use of these recurrence equations, these polynomials are expressible through additional relations. First, as vectors $|n\rangle$ represent an orthogonal normalized system, we write

$$(n|m) = \sum_{\beta\gamma}\langle n, \beta|m, \gamma\rangle = \delta_{nm};$$

hence,

$$\sum_{(\beta\gamma)\alpha} \Pi_{\beta\gamma}(n, m) = 0, \quad \alpha > 0, \tag{3.7}$$

in which $\Pi_{\alpha\beta}(n, m) = \Pi^0_{\alpha\beta}(n, m)$. Eq. (3.7) provides the normalization of the wavefunction in the perturbation theory of any order. Second, through condition $\sum_\ell |\ell\rangle\langle\ell| = 1$, we have

$$\langle n, \alpha|\xi^{s+q}|m, \beta\rangle = \sum_\ell \langle n, \alpha|\xi^s|\ell\rangle\langle\ell|\xi^q|m, \beta\rangle.$$

Then, on converting to polynomials, we obtain this addition theorem:

$$\Pi^{s+q}_{\alpha\beta}(n, m) = \sum_{\ell < n} g_{\ell n}\Pi^s_{0\alpha}(\ell, n)\Pi^q_{0\beta}(\ell, m) + \sum_{n \leq \ell \leq m} \Pi^s_{\alpha 0}(n, \ell)\Pi^q_{0\beta}(\ell, m)$$
$$+ \sum_{\ell > m} g_{m\ell}\Pi^s_{\alpha 0}(n, \ell)\Pi^q_{\beta 0}(m, \ell).$$

Within the formalism under consideration, an important role is evidently played, as it must be, by polynomials $\Pi_{0\alpha}(n - k, n)$ and $\Pi_{\alpha 0}(n, n + k)$ and, of course, harmonic polynomials $\Pi^s(n, m) = \Pi^s_{00}(n, m)$.

The initial data are harmonic polynomials $\Pi^s(n, m)$, which are found according to the addition theorem or trivial calculations of matrix elements $\langle n|\xi^s|m\rangle$:

$$\Pi^s(n, n + s) = 1, \quad \Pi^2(n, n) = 2n + 1, \quad \text{etc.}$$

The first polynomials are linear in a_1 and follow from the recurrence equations. For instance,

$$\Pi^3_{01}(n, n) = \Pi^3_{10}(n, n) = -a_1(30n^2 + 30n + 11).$$

The *convolution* $\Pi^3_{(\beta\gamma)1}(0)$ together with $\Pi^4(n,n)$ clearly forms the energy in the second-order approximation. By definition, the operation of convolution or contraction of polynomials represents the sum of polynomials of the same order:

$$\Pi^s_{(\beta\gamma)\alpha}(+k) = \sum_{(\beta\gamma)\alpha} \Pi^s_{\beta\gamma}(n, n+k) \text{ and } \Pi^s_{(\beta\gamma)\alpha}(-k) = \sum_{(\beta\gamma)\alpha} \Pi^s_{\beta\gamma}(n-k, n);$$

the result of convolution invariably yields a degree decreased from the highest degree of the convolving polynomials. Second-order polynomials are determined in an analogous manner. We initially calculate Π_{02}; from identity $\Pi_{(\beta\gamma)2}(\pm k) = 0$, we then find Π_{20} and, eventually with the addition theorem, we reconstruct Π^s_{02} and Π^s_{20}. This procedure is repeated for approximations of third order and greater orders.

We discuss a numerical value of the highest degree in quantum number of the polynomial.[8] After quantization, variable ξ practically converts into $\sqrt{n}$, so that $\xi^s \to n^{s/2}$. Vector $|n, \alpha\rangle$ is formed primarily by quantity $\xi^{3\alpha}$, and $|n+k, \beta\rangle$ is formed analogously by $\xi^{3\beta}$; therefore,

$$\Pi^s_{\alpha\beta}(n, n+k) \sim n^{(s+3\alpha+3\beta-k)/2},$$

because

$$\Pi^s_{\alpha\beta}(n, n+k) = \left(\sqrt{g_{n,n+k}}\right)^{-1} \langle n, \alpha | \xi^s | n+k, \beta \rangle.$$

Taking into account that $3(\alpha + \beta) + s = k_{\max}$, we find for the highest degree in n of both polynomials $\Pi^s_{\alpha\beta}(n, n+k)$ and $\Pi^s_{\beta\alpha}(n-k, n)$ the value $(k_{\max} - k)/2$ with $k \neq 0$. If $k = 0$ and $\alpha + \beta$ is an odd number, then the resultant degree decreases by unity and becomes equal to $(k_{\max}/2) - 1$. We arrive at this conclusion readily if we take into account, for example, that $k_{\max} \geq 2$ in this case. For an even value of $\alpha + \beta$, when $k = 0$, the highest degree in n is simply equal to $k_{\max}/2$.

As an illustration, we write two polynomials

$$\Pi^3_{02}(n, n+5) = \frac{a_1^2}{12}(80n^2 + 495n + 639) + \frac{a_2}{4}(11n + 39)$$

and

$$\Pi^3_{20}(n, n+5) = \frac{a_1^2}{12}(80n^2 + 465n + 549) - \frac{a_2}{4}(11n + 27),$$

add to them this one

$$\Pi^3_{11}(n, n+5) = -\frac{a_1^2}{3}(40n^2 + 240n + 234),$$

and calculate the convolution $\Pi^3_{(\beta\gamma)2}(5)$. Clearly, $\Pi^3_{(\beta\gamma)2}(5) = 21a_1^2 + 3a_2$; the highest degree $(k_{max} - k)/2$ here is equal to 2 decreases, and the sought result becomes much simpler.

We proceed to formulate the principal definitions in terms of the polynomial language. It is convenient to suppose that $\lambda = 1$; in this case, coefficients a_p pertain to a small order, i.e., $a_p \sim \lambda^p$. Furthermore, only in exceptional cases do we show parameter λ explicitly. The energy of anharmonic vibrations in the one-dimensional case becomes written as

$$E_n = \hbar\omega\left(n + \frac{1}{2}\right) + \sum_\alpha E_n^\alpha, \quad E_n^\alpha = \frac{\hbar\omega}{\alpha} \sum_{(p\beta\gamma)\alpha} pa_p\Pi_{\beta\gamma}^{p+2}(n,n).$$

The summation is clearly taken solely with respect to even values of α because only even corrections to the vibrational energy are nonzero. This circumstance is important. The first-order correction to the energy is equal to zero; the second-order correction is proportional to a_1^2 and a_2, and the next nonvanishing correction is linear in a_1^4, $a_1^2 a_2$, a_2^2, $a_1 a_3$, and a_4, i.e., the correction proportional to a_1^3, $a_1 a_2$, and a_3 vanishes. Continuing in such a manner, we obtain the exact relations for corrections of higher order in complete agreement with the experimental data.

The arbitrary correction to the wavefunction takes a simple form,

$$|n,\alpha\rangle = \sum_{k=0}^{3\alpha} \sqrt{g_{n-k,n}}\Pi_{0\alpha}(n-k,n)|n-k\rangle + \sum_{k=1}^{3\alpha} \sqrt{g_{n,n+k}}\Pi_{\alpha0}(n,n+k)|n+k\rangle,$$

in which the summation is extended over values of k with parity identical to that of α. Despite the apparent triviality of this expression for $|n,\alpha\rangle$, the wavefunction has a latent role in the polynomial formalism.

Finally, we consider a function $f(\xi)$, which is expanded as a power series in ξ:

$$f(\xi) = \sum_s \frac{2^{-s/2}}{s!} f^{(s)}\xi^s.$$

Derivatives $f^{(s)}$ are chosen so that this expansion is a Taylor series for variable q; $\sqrt{2}q = \xi$. With $F_s = 2^{-s/2}f^{(s)}/s!$, the matrix element becomes expressed as

$$\langle n|f|n+k\rangle = \sum_{s\beta\gamma} F_s\langle n,\beta|\xi^s|n+k,\gamma\rangle.$$

From this formula and Eq. (3.6), we obtain

$$\langle n|f|n+k\rangle = \sqrt{g_{n,n+k}} \sum_{s\alpha} \sum_{(\beta\gamma)\alpha} F_s\Pi_{\beta\gamma}^s(n,n+k). \tag{3.8}$$

This general scheme to construct the new formalism is sufficiently simple. The recurrence equations thus derived enable a definition of an arbitrary polynomial in an explicit form. By this means, we calculate all desired polynomials; the sought expressions, in particular, for the eigenvalues and eigenfunctions, then depend only on the accuracy of the required approximation.

Ensemble of anharmonic oscillators

The calculations in perturbation theory for a system with variables of arbitrary number differ substantially from those for a one-dimensional case, even in the first order. The calculation of the matrix elements, for instance, with the help of Eq. (3.3), is accompanied by competition among various mechanical approximations and yields cumbersome expressions. Furthermore, the derivatives of the dipolar moment that have a maximum influence on the matrix elements are unknown. Using the polynomial technique, we consider these questions in detail.

Returning to Eq. (3.1), we consider an arbitrary correction $|n, \alpha\rangle$ to the function in the harmonic approximation. Vector $|n, \alpha\rangle$ is constructed from all possible states $|n_1 \pm \ell_1, \ldots, n_i \pm \ell_i, \ldots, n_r \pm \ell_r\rangle$, in which $\ell_1 + \ell_2 + \cdots + \ell_r \leq 3\alpha$. Functions $|n_i + \ell_i\rangle$ and $|n_i - \ell_i\rangle$ are multiplied by factors $\sqrt{g_{n_i, n_i + \ell_i}}$ and $\sqrt{g_{n_i - \ell_i, n_i}}$, respectively. Recall that state vector $|n_1, n_2, \ldots, n_r\rangle$ is a product of individual functions $|n_1\rangle, |n_2\rangle, \ldots, |n_r\rangle$. We construct an arbitrary matrix element between separate corrections $|n, \alpha\rangle$ and $|n + k, \beta\rangle$:

$$\langle n, \alpha | \xi_1^{s_1} \xi_2^{s_2} \ldots \xi_r^{s_r} | n + k, \beta \rangle.$$

This matrix element is determined essentially by elements $\langle n_i \pm \ell_i | \xi_i^{s_i} | n_i \pm p_i + k_i \rangle$:

$$\sqrt{g_{n_i, n_i + \ell_i} g_{n_i + k_i, n_i + p_i + k_i}} \langle n_i + \ell_i | \xi_i^{s_i} | n_i + p_i + k_i \rangle,$$

$$\sqrt{g_{n_i - \ell_i, n_i} g_{n_i + k_i, n_i + p_i + k_i}} \langle n_i - \ell_i | \xi_i^{s_i} | n_i + p_i + k_i \rangle, \text{ etc.}$$

According to relation $\langle \ell | \xi^q | \ell + p \rangle \sim \sqrt{g_{\ell, \ell + p}}$, these matrix elements are proportional to factor $\sqrt{g_{n_i, n_i + k_i}}$; hence,

$$\langle n, \alpha | \xi_1^{s_1} \xi_2^{s_2} \ldots \xi_r^{s_r} | n + k, \beta \rangle = \sqrt{g_{n, n+k}} \Pi_{\alpha\beta}^s(n, n+k),$$

in which the polynomial

$$\Pi_{\alpha\beta}^{s_1 s_2 \ldots s_r}(n_1, n_1 + k_1; \ n_2, n_2 + k_2; \ \ldots; \ n_r, n_r + k_r)$$

of quantum numbers $n_1, n_2, \ldots n_r$ is designated as $\Pi_{\alpha\beta}^s(n, n+k)$. Moreover, here, $g_{n,n+k} \equiv g_{n_1, n_1 + k_1} g_{n_2, n_2 + k_2} \cdots g_{n_r, n_r + k_r}$, and sets $\{s_1, s_2, \ldots, s_r\}$ and $\{k_1, k_2, \ldots, k_r\}$ are denoted by s and k, respectively.

This definition of polynomials produces a general selection rule for k_i. Matrix element $\langle n, \alpha | \xi_1^{s_1} \xi_2^{s_2} \ldots \xi_r^{s_r} | n + k, \beta \rangle$ is determined by a sum of matrix elements $\langle n_i \pm \ell_i | n_i \pm p_i \pm c_{si} + k_i \rangle$ in various combinations, in which c_{si} has the parity of number s_i; equality $k_i = \pm \ell_i \pm p_i \pm c_{si}$ therefore becomes satisfied. On summing this equality with respect to i, we obtain the desired rule,

$$\sum_i k_i \leftrightarrow \alpha + \beta + \sum_i s_i,$$

which asserts that polynomials $\Pi_{\beta\alpha}^s(n-k,n)$ and $\Pi_{\alpha\beta}^s(n,n+k)$ differ from zero only for a case in which $\sum_i k_i$ has the parity of number $\alpha + \beta + \sum_i s_i$.

By analogy with the one-dimensional case, from Eq. (3.1), we convert from the matrix elements to polynomials and derive the corresponding recurrence equations,

$$\Pi_{\mu\alpha}^s(\ell,n) = \frac{1}{\alpha} \sum_{(pq\beta\gamma\nu)\alpha} p \sum_{(j)p+2} a_j \left[\sum_{m<\ell} g_{m\ell} \Delta_q \Pi_{\nu\mu}^s(m,\ell) \Pi_{\beta\gamma}^j(m,n) \right.$$
$$\left. + \sum_{\ell \leq m < n} \Delta_q \Pi_{\mu\nu}^s(\ell,m) \Pi_{\beta\gamma}^j(m,n) + \sum_{m>n} g_{nm} \Delta_q \Pi_{\mu\nu}^s(\ell,m) \Pi_{\gamma\beta}^j(n,m) \right],$$

$$\Pi_{\alpha\mu}^s(n,\ell) = \frac{1}{\alpha} \sum_{(pq\beta\gamma\nu)\alpha} p \sum_{(j)p+2} a_j \left[\sum_{m<n} g_{mn} \Delta_q \Pi_{\nu\mu}^s(m,\ell) \Pi_{\beta\gamma}^j(m,n) \right.$$
$$\left. + \sum_{n<m\leq \ell} \Delta_q \Pi_{\nu\mu}^s(m,\ell) \Pi_{\gamma\beta}^j(n,m) + \sum_{m>\ell} g_{\ell m} \Delta_q \Pi_{\mu\nu}^s(\ell,m) \Pi_{\gamma\beta}^j(n,m) \right],$$

in which j implies a set $\{j_1, j_2, \ldots, j_r\}$, and factor $\Delta_q(n,m)$ is given by expression

$$\sum_i \sum_{(\alpha_1\alpha_2\ldots\alpha_i)q} \frac{(E_m^{\alpha_1} - E_n^{\alpha_1})(E_m^{\alpha_2} - E_n^{\alpha_2})\ldots(E_m^{\alpha_i} - E_n^{\alpha_i})}{(E_n^0 - E_m^0)^{i+1}},$$

with $\Delta_0(n,m) = (E_n^0 - E_m^0)^{-1}$. Inspection of the first summation in the recurrence relations shows that, in accordance with the equality $p + q + \beta + \gamma + \nu = \alpha$, the indices q, β, γ, and ν take values from 0 to $\alpha - 1$, whereas $p = 1, 2, \ldots, \alpha$. The subscripts of the polynomials on the right side apparently do not exceed $\alpha - 1$. Because the subscripts determine the order of the polynomial in λ, we obtain the solution for arbitrary polynomials $\Pi_{\alpha\mu}^s(n, n \pm k)$ beginning with $\alpha = 1$ and $\mu = 0$. The quantities E_n^α appearing in the factor $\Delta_q(n,m)$ are corrections of order λ^α to harmonic oscillator energy E_n^0;

$$E_n^0 = \hbar \sum_{i=1}^r \omega_i \left(n_i + \frac{1}{2} \right),$$

in which ω_i are the frequencies of harmonic vibrations. The denominator of the expression for $\Delta_q(n,m)$ contains only differences $E_n^0 - E_m^0$ to varied degrees, which evidently produce no dependence on n after summation over m. This condition proves that the quantities in question are polynomials of quantum numbers $n_1, n_2, \ldots, n_r$.

The polynomials are symmetric under simultaneous permutations of the subscripts and all pairs of quantum numbers:

$$\Pi^s_{\alpha\beta}(n, n \pm k) = \Pi^s_{\beta\alpha}(n \pm k, n).$$

The polynomials with zero subscripts $\Pi^s(n, n \pm k)$ are readily calculated with the aid of the addition theorem or simply through the matrix element. It is convenient to tabulate not the polynomials themselves but rather their convolutions, i.e., the sums of polynomials of the same order:

$$\Pi^s_{(\beta\gamma)\alpha}(\pm k) \equiv \sum_{(\beta\gamma)\alpha} \Pi^s_{\beta\gamma}(n, n \pm k).$$

The convolution operation decreases the highest degree in n and is applied to polynomials $\Pi^s_{\beta\gamma}(n_1, n_1 + k_1; n_2 - k_2, n_2; n_3, n_3 - k_3, \ldots)$ with arbitrary values of k. Note here that, invariably,

$$k_i \le k_{\max} = 3(\beta + \gamma) + \sum_i s_i,$$

and the polynomials vanish for $k_i > k_{\max}$.

Having applied this correspondence, we write the exact energy of anharmonic vibrations as

$$E_n = \hbar \sum_{i=1}^{r} \omega_i \left(n_i + \frac{1}{2}\right) + \sum_\alpha E_n^\alpha, \; E_n^\alpha = \frac{1}{\alpha} \sum_{(p\beta\gamma)\alpha} p \sum_{(j)p+2} a_j \Pi^j_{\beta\gamma}(n,n).$$

In terms of polynomials, the arbitrary correction to wavefunction $|n, \alpha\rangle$, the polynomial addition theorem, and Eq. (3.7) that imposes normalization of the wavefunction remain valid and, moreover, exactly retain their form. In the final expressions, $\lambda = 1$.

We "translate" the formula for matrix elements of an arbitrary coordinate function

$$f = \sum_\ell \sum_{(s)\ell} \frac{2^{-\ell/2}}{\ell!} f^{(\ell)}_{s_1 s_2 \ldots s_r} \xi_1^{s_1} \xi_2^{s_2} \ldots \xi_r^{s_r}.$$

Here, $f_s^{(\ell)}$ are ordinary derivatives in a Taylor series expansion of function f in normal coordinates q_i. If $|n\rangle = \sum_\alpha |n, \alpha\rangle$, in which $|n, 0\rangle$ is the harmonic state vector, then

$$\langle n|f|n+k\rangle = \sum_\ell \sum_{(s)\ell} \frac{2^{-\ell/2}}{\ell!} f^{(\ell)}_{s_1 s_2 \ldots s_r} \sum_{\alpha\beta} \langle n, \alpha|\xi_1^{s_1} \xi_2^{s_2} \ldots \xi_r^{s_r}|n+k, \beta\rangle.$$

Having used the polynomial definition, we derive formula

$$(n|f|n+k) = \sqrt{g_{n,n+k}} \sum_\ell \sum_{(s)\ell} \frac{2^{-\ell/2}}{\ell!} f_s^{(\ell)} \sum_{\alpha\beta} \Pi^s_{\alpha\beta}(n, n+k),$$

which has an obvious coincidence with Eq. (3.8).

General equations

We proceed to apply the results, which are obtained in a framework of the solution of the general problem of degenerate states, to the problem of anharmonicity of interest. In this case,

$$G_p = \sum_{(j_1 j_2 \ldots j_r) p + 2} a_{j_1 j_2 \ldots j_r} \xi_1^{j_1} \xi_2^{j_2} \ldots \xi_r^{j_r},$$

the eigenvalues of operator H^0 are

$$E_n^0 = \hbar \sum_{i=1}^r \omega_i \left(n_i + \frac{1}{2} \right),$$

and the expansions

$$|nx\rangle = \sum_{x_1 x_2 \ldots x_r} U^n_{xx_i} |n_1 + x_1, \ldots, n_i + x_i, \ldots, n_r + x_r\rangle$$

in the harmonic oscillator eigenvectors

$$|n_1, n_2, \ldots, n_r\rangle = |n_1\rangle|n_2\rangle \ldots |n_r\rangle$$

with the coefficients of an appropriate unitary transformation $U^n_{xx_i}$ should be chosen as the correct functions for initial states. Here, x_i are known integers (both positive and negative) that specify the complete set of degenerate vibrational states of level n and are determined from the conditions

$$\sum_{i=1}^r \omega_i x_i = 0.$$

Here, ω_i are the frequencies of harmonic vibrations.

We assume the vibrational levels to be degenerate; consequently, $\sigma > 0$. Then, according to Eq. (3.4),

$$|nx, 1\rangle = \sum_{y; m \neq n} \sum_{(j_1 \ldots j_r) 3} a_{j_1 j_2 \ldots j_r} \frac{\langle my|\xi_1^{j_1} \xi_2^{j_2} \ldots \xi_r^{j_r}|nx\rangle}{E_n^0 - E_m^0} |my\rangle.$$

Because $\sum_y (U_{yy_i}^m)^* U_{yy_i'}^m = \delta_{y_i y_i'}$, correction $|nx, 1\rangle$ is an expansion in these vectors:

$$\prod_{i=1}^r \sqrt{g_{n_i+x_i, n_i+x_i \pm \ell_i}} |n_i + x_i \pm \ell_i\rangle, \text{ in which } \ell_1 + \ell_2 + \cdots + \ell_r \leq 3.$$

Correction $|nx, 2\rangle$ is represented with a similar expansion but with $\ell_1 + \ell_2 + \cdots + \ell_r \leq 6$. Finally, it is easily shown by induction that, in the general case, correction $|nx, \alpha\rangle$ is expanded in the functions

$$\prod_{i=1}^r \sqrt{g_{n_i+x_i, n_i+x_i \pm \ell_i}} |n_i + x_i \pm \ell_i\rangle, \quad \text{in which } \ell_1 + \ell_2 + \cdots + \ell_r \leq 3\alpha;$$

each correction contains coefficient $U_{xx_i}^n$ appearing necessarily on summation over x_i. We introduce the matrix element,

$$M_{\alpha\beta}^{s_1 s_2 \ldots s_r}(nx, my) = \langle nx, \alpha | \xi_1^{s_1} \xi_2^{s_2} \ldots \xi_r^{s_r} | my, \beta \rangle.$$

In view of the properties of corrections $|nx, \alpha\rangle$ and $|my, \beta\rangle$ considered here, $M_{\alpha\beta}^s(nx, my)$ clearly comprises various elements $\langle n_i + x_i \pm \ell_i | \xi_i^{s_i} | m_i + y_i \pm \ell_i' \rangle$ with the corresponding factors $\sqrt{g_{n_i+x_i, n_i+x_i \pm \ell_i}}$ and $\sqrt{g_{m_i+y_i, m_i+y_i \pm \ell_i'}}$, but $\langle k | \xi^q | p \rangle \sim \sqrt{g_{kp}}$; consequently, in a manner analogous with the nondegenerate case, one might introduce the polynomial structures

$$M_{\alpha\beta}^s(nx, my) = \sum_{x_i y_i} \sqrt{g_{n_1+x_1, m_1+y_1} g_{n_2+x_2, m_2+y_2} \cdots g_{n_r+x_r, m_r+y_r}} \Pi_{\alpha\beta}^s(nxx_i, myy_i).$$

Quantities $\Pi_{\alpha\beta}^s(nxx_i, myy_i)$ appear to be polynomials only from a computational point of view. The expressions following from this perturbation theory have a polynomial form after the calculation of the corresponding matrix elements; through the initial coefficients $U_{xx_i}^n$ and because, within the group of degenerate states, we must retain $E_{nx}^\sigma - E_{ny}^\sigma$ rather than $E_{nx}^0 - E_{ny}^0$ in the denominators of expansions, the dependences of $\Pi_{\alpha\beta}^s(nxx_i, myy_i)$ on the quantum numbers might be more complicated than merely of polynomial form. For this reason, it is preferable to derive the principal recurrence relations not for polynomials but rather for matrix elements $M_{\alpha\beta}^s(nx, my)$, from which the polynomial structures become readily reconstructed according to the aforementioned definition.

With regard to the equations for matrix elements, it suffices to multiply Eq. (3.4) for $|nx, \alpha\rangle$ by $\langle \ell z, \mu | \xi_1^{s_1} \xi_2^{s_2} \ldots \xi_r^{s_r}$ on the left. As a result, we obtain

$$M_{\mu\alpha}^s(\ell z, nx) = \frac{1}{\alpha} \sum_{y; m \neq n} \sum_{(pq\beta\gamma\nu)\alpha} p \sum_{(j)p+2} a_j \Delta_q^0(nx, my) M_{\mu\nu}^s(\ell z, my) M_{\beta\gamma}^j(my, nx)$$

$$+ \frac{1}{\alpha} \sum_{y \neq x} \sum_{(pq\beta\gamma\nu)\alpha > \sigma} p \sum_{(j)p+2} a_j \Delta_q^\sigma(nx, ny) M_{\mu\nu}^s(\ell z, ny) M_{\beta\gamma}^j(ny, nx),$$

(3.9)

in which indices j and s denote, as before, integers in sets $\{j_1, j_2, \ldots, j_r\}$ and $\{s_1, s_2, \ldots, s_r\}$, respectively. These general recurrence relations determine an arbitrary matrix element in the presence of degeneracy. The nondegenerate case is clearly contained here for $\sigma = 0$. Thus, one can reconstruct all elements $M^s_{\alpha\beta}(nx, my)$ and, assuming $\lambda = 1$ in the final formulae, calculate the eigenvalues

$$E_{nx} = E_n^0 + \sum_\alpha \frac{1}{\alpha} \sum_{(p\beta\gamma)\alpha} p \sum_{(j)p+2} a_j M^j_{\beta\gamma}(nx, nx)$$

of an anharmonic Hamiltonian H and the matrix elements

$$(nx|f|my) = \sum_{\ell\beta\gamma} \sum_{(s)\ell} \frac{2^{-\ell/2}}{\ell!} f_s^{(\ell)} M^s_{\beta\gamma}(nx, my)$$

of a function f, of which the explicit form is determined according to the expression,

$$(nx|f|my) = \sum_{x_i y_i} \sqrt{g_{n_1+x_1, m_1+y_1} g_{n_2+x_2, m_2+y_2} \cdots g_{n_r+x_r, m_r+y_r}} \Phi(nxx_i, myy_i);$$

$\Phi(nxx_i, myy_i)$ are, in turn, functions of quantum numbers $n_1, n_2, \ldots, n_r$ and $m_1, m_2, \ldots, m_r$. In addition, Φ depends trivially on numbers x_i and y_i, which are zero in the absence of degeneracy ($x_i = y_i = 0$).

In the general case, to calculate the observable quantities, one must use Eq. (3.9), which allows one to take into account the degenerate levels. From a practical point of view, for both one-dimensional and many-dimensional problems, Eq. (3.9) is principal; having obtained with its aid all matrix elements and with the result divided by factor $\sqrt{g}$, we reconstruct polynomial structures. One should express matrix elements in a polynomial manner and tabulate not the polynomials but rather their convolutions. Although we can work with equations of polynomials, Eq. (3.9) is convenient and simple for the calculation of higher order approximations in the perturbation theory. The equations with polynomials are necessary to exhibit and to prove the polynomial structure of Π quantities in explicit form.

The problem of general equations is nearly solved; the selection rule remains. Quantities $\Pi^s_{\alpha\beta}(nxx_i, myy_i)$ are nonzero provided that the parities of $\sum_i[(n_i + x_i) - (m_i + y_i) + s_i]$ and $\alpha + \beta$ coincide. This condition follows from the selection rule for the polynomials in the nondegenerate case on replacements $n_i \to n_i + x_i$ and $m_i \to m_i + y_i$.

Physical interpretation

The eigenvalues of the Hamiltonian determine the frequencies, i.e., the eigenvectors — the intensities of quantum transitions. With intensities of electromagnetic radiation I_0

before and I after its travel through a layer of matter of thickness b with concentration C of molecules, according to the Bouguer–Lambert–Beer law,

$$I = I_0 e^{-\alpha(\omega)bC},$$

in which coefficient $\alpha(\omega)$ characterizes the ability of matter to absorb the radiation of frequency ω. Quantity $\alpha(\omega)$ is generally represented as

$$\alpha(\omega) = S_{mn} f(\omega - \omega_0),$$

in which S_{mn} is the line strength and $f(\omega - \omega_0)$ is a function that defines a shape or contour of a spectral line with center ω_0.[9,10] Line strength S_{mn} is proportional to the transition energy, $\hbar\omega_{mn}$; apart from that common factor, the type of absorption is determined by the Einstein probabilities, i.e., by the squared matrix elements of electric $\mathbf{d}$ and magnetic $\boldsymbol{\mu}$ dipolar moments, electric quadrupolar moment Θ, and so on. Essentially, the quantity S_{mn} defines the intensity. We have

$$S_{mn} = \hbar\omega_{mn}(c_1|\mathbf{d}_{mn}|^2 + c_2|\boldsymbol{\mu}_{mn}|^2 + c_3|\Theta_{mn}|^2 + \cdots).$$

Coefficients $c_1, c_2, c_3, \ldots$ are independently determined for each concrete case of a physical problem.

From a practical point of view, the electric dipolar transitions are of most interest. For example, consider this expression for S_{mn} for free molecules:

$$S_{mn} = \hbar\omega_{mn} \cdot \frac{4\pi^2}{3\hbar c} |(m|\mathbf{d}|n)|^2 (1 - e^{-\hbar\omega_{mn}/k_B T})(Ng_n/Q)e^{-E_n/k_B T}.$$

Here, all quantities are simply interpreted.[9,10] Specifically,

$$\frac{4\pi^2}{3\hbar c} |(m|\mathbf{d}|n)|^2$$

follows from an expression for the transition probability per second obtained in the first order of perturbation theory. Transition energy $\hbar\omega_{mn}$ is equal to $E_m - E_n$, of which E_n and E_m are the energies of a molecule belonging to eigenstates $|n\rangle$ and $|m\rangle$. Quantity

$$N_n = (Ng_n/Q)e^{-E_n/k_B T}$$

from the Boltzmann law defines a fraction of molecules in the initial state with energy E_n at temperature T. Here, N is the concentration of molecules, k_B is the Boltzmann constant, g_n is the degeneracy of level E_n, and Q is the partition function, for which

$$Q = \sum_s g_s e^{-E_s/k_B T}.$$

The factor

$$1 - e^{-\hbar\omega_{mn}/k_B T} = 1 - N_m g_n / N_n g_m,$$

in which N_m is a number of molecules in the final state with degeneracy g_m, takes into account the effects of induced emission. This factor is generally near unity.

The principal problem of a calculation of intensity is reduced to the calculation of matrix elements of the electric dipolar moment function between exact eigenfunctions of a molecular Hamiltonian. This procedure is generally laborious. The problem of a calculation of matrix elements is further complicated in that a correct explanation of spectra must take into account the anharmonicity caused by the nonlinearity of the dipolar moment function d. So, for an arbitrary polyatomic molecule, we have

$$d = \sum_\ell \sum_{(s)\ell} \frac{2^{-\ell/2}}{\ell!} d^{(\ell)}_{s_1 s_2 \ldots s_r} \xi_1^{s_1} \xi_2^{s_2} \ldots \xi_r^{s_r}.$$

Coefficients $d^{(\ell)}_{s_1 s_2 \ldots s_r}$ in this expansion of the dipolar moment function in normal coordinates q_k characterize the *electro-optical* anharmonicity of molecular vibrations:

$$q_k = \frac{\xi_k}{\sqrt{2}}.$$

The higher the overtone, the greater the influence of the nonlinear part of function d on overtone intensity.

Rules for polynomials

First, the polynomials form, with the required accuracy, all necessary physical observables of the anharmonicity problem. The desired quantities are obtained immediately on solving or opening the recurrence equations or relations avoiding conventional intermediate manipulations. We compare two schemes to construct the stationary perturbation theory:

1. Schrödinger equation → eigenfunctions and eigenvalues → matrix elements
   and
2. recurrence equations → eigenvalues and matrix elements.

The first scheme is conventional; however, we proposed the second scheme. The main disadvantage of the conventional scheme is that, at each stage, one must return virtually to the beginning — to the Schrödinger equation — to improve the eigenfunctions by increasing the order of the perturbation calculation. Only after these calculations is one in a position to evaluate the matrix elements. In our method, intermediate calculations are performed on an equal footing, i.e., the procedures to calculate the eigenvalues and arbitrary matrix elements are performed simultaneously.

Second, the proposed theory automatically keeps track of nonzero contributions of the total perturbation to the result sought (see the selection rule discussed later) and takes into account the history of the calculations, i.e., the intermediate calculations. This advantage is achieved on expanding, in a small parameter, the derivatives of the energies and their wavefunctions rather than by expanding the eigenfunctions and eigenvalues, as is traditionally performed. In this sense, the expansion in exact eigenvectors plays a principal role,

$$\frac{\partial}{\partial \lambda}|n, \lambda\rangle = \sum_{m \neq n} C_{mn}|m, \lambda\rangle,$$

because it ensures the full use of the history of the calculations and, consequently, significantly simplifies the general solution algorithm. If the expansion is performed in terms of the exact eigenvectors, rather than in terms of zero-order basis functions, then it is assumed that the former functions exist and are expressible algebraically, for example, with recurrence relations. In addition, one might avoid the renormalization of the function; this problem presents considerable difficulties in the traditional approach, in which the function should be renormalized on passing from one perturbation order to the next.

Other advantages of this method appear in various applications of this perturbation theory.[5,7,8] For example, in a framework of the polynomial formalism, one might consider the problem of electro-optical anharmonicity; this problem involves an electric dipolar moment function d in a nonlinear form, and its solution requires evaluation of matrix elements $(n|d|m)$. The absolute values of dipolar moment derivatives $d^{(s)}$ might be unknown beforehand, which complicates the problem. In the traditional formalism, the consideration proceeds, as a rule, from the wavefunction of a definite order, which leads to the loss of significant contributions. In the polynomial formalism, we separately consider each term in an expansion of the dipolar moment function and, consequently, calculate the entire matrix element in a given order in a small parameter. For instance, let

$$d(q) = d^0 + d'q + \frac{d''}{2}q^2 + \frac{d'''}{6}q^3$$

be a model dipolar moment function that depends on only one normal coordinate q. If the anharmonicity is such that d'' is ~ 10 times d' and ~ 100 times d''', then a conventional calculation of matrix element $(n|d|m)$ in the second order of perturbation theory yields

$$(n|d|m) = d^0 \delta_{nm} + \underbrace{d'(n|q|m) + \frac{d''}{2}(n|q^2|m) + \frac{d'''}{6}(n|q^3|m)}_{\text{second order}}.$$

This result is incorrect, however, because the matrix element is a sum of terms of disparate orders. Namely, $d'(n|q|m)$ is a third-order quantity and $d'''(n|q^3|m)$ is a fourth-order quantity. To improve this situation, one must calculate in a somewhat different manner (see the rule of order):

$$(n|d|m) = d^0 \delta_{nm} + \underbrace{d'(n|q|m)}_{\text{first order}} + \underbrace{\frac{d''}{2}(n|q^2|m)}_{\text{second order}} + \underbrace{\frac{d'''}{6}(n|q^3|m)}_{\text{zero order}}.$$

This approach is especially simple to implement in a polynomial formalism. It is necessary in the above case to evaluate two convolutions, $\Pi^2_{(\alpha\beta)2}$ and $\Pi^1_{(\alpha\beta)1}$, and to reconstruct the harmonic polynomial, Π^3.

Summarizing the above analysis, the observable intensities and frequencies of quantum transitions are associated physically with matrix elements. Frequencies are associated with differences of diagonal matrix elements of a Hamiltonian but intensities with matrix elements of dipolar moment function d,

$$(n|d|n+k) = \sqrt{g_{n,n+k}} \sum_{\ell\alpha} \sum_{(s_1 s_2 \ldots s_r)\ell} \frac{2^{-\ell/2}}{\ell!} d^{(\ell)}_{s_1 s_2 \ldots s_r} \Pi^{s_1 s_2 \ldots s_r}_{(\beta\gamma)\alpha}(k).$$

Because the quantum mechanical amplitude λ is typically $\sim 10^{-1}$, the expansion coefficients of the dipolar moment function $d^{(\ell)}_s$ can be assumed to be proportional to λ^{σ_s}, in which $\sigma_{s_1 s_2 \ldots s_r}$ is an integer that determines the order of $d^{(\ell)}_{s_1 s_2 \ldots s_r}$ in terms of λ. If electro-optical effects are weak, then $\sigma_s = \ell$; hence, the difference between σ_s and ℓ characterizes the strength of electro-optical anharmonicity. This condition becomes a definition of electro-optics. Expanding the dipolar moment function in terms of vibrational variables,

$$d = \sum_\ell \sum_{(s)\ell} \frac{2^{-\ell/2}}{\ell!} d^{(\ell)}_{s_1 s_2 \ldots s_r} \xi^{s_1}_1 \xi^{s_2}_2 \ldots \xi^{s_r}_r,$$

we obtain automatically the dependence of the matrix element on the quantum mechanical amplitude. Quantity $\xi^{s_i}_i$ is associated with λ^{s_i}; however, although it might seem that $d^{(\ell)}_s \sim \lambda^\ell$, this association is incorrect. The behavior of derivatives $d^{(\ell)}_s$ can deviate strongly from that of λ^ℓ, which indicates the presence of another (electro-optical) nature of anharmonicity, as distinct from the mechanical anharmonicity related to the Hamiltonian. The greater the difference between σ_s and ℓ, the stronger the electro-optical anharmonicity; the equality $\sigma_s = \ell$ is indicative of the absence of the latter. One might apply this analysis to an arbitrary coordinate function f, of physical interest, in an analogous manner.

The general rules pertinent for calculating the matrix elements are:

1. for the matrix element to be represented in the same order of a small parameter, it suffices to satisfy the equality $\sigma_{s_1 s_2 \ldots s_r} + \alpha = \text{const}$ (the rule of order) and

2. for convolutions $\Pi^{s_1 s_2 \ldots s_r}_{(\beta\gamma)\alpha}(k)$ to be nonzero, numbers $\sum_i (k_i - s_i)$ and α need to have the same parity; otherwise, the polynomials vanish identically (the selection rule).

With respect to the rule of order, some comments have been discussed; here, we consider the selection rule. Let the value of sum $\sum_i k_i$ be odd; the contributions with even ℓ should then be taken into account for odd values of α, and the contributions with odd values of ℓ are associated with even values of α. However, if the value of $\sum_i k_i$ is even, then both ℓ and α are either even or odd. Through the selection rule, half of all possible contributions of the perturbation to an arbitrary matrix element vanish. The same condition is true for eigenvalues E_n. As $E_n \sim \Pi^{j_1 j_2 \ldots j_r}_{(\beta\gamma)\nu}(0)$, the sum of all j_i and ν should be even; the other variants result in zero. This polynomial technique possesses a pronounced structure: all necessary quantities are directly determined in terms of nonzero polynomials or through their convolutions, which can be tabulated to facilitate calculations.

Quantum functions

A prospectively useful direction for further investigation is to proceed beyond solutions with the perturbation theory. We assume that the effective internuclear potential is a real function that is represented as an expansion in a power series in terms of the normal coordinates. In this case, the procedure of quantization, i.e., the calculation of matrix elements of an arbitrary coordinate function, taking into account the influence of anharmonicity, is reduced to the sum of polynomials multiplied by factor $\sqrt{g}$:

$$\langle n|f|n+k\rangle = \sqrt{g_{n,n+k}} \sum_{\ell\alpha} \sum_{(s)\ell} \frac{2^{-\ell/2}}{\ell!} f_s^{(\ell)} \Pi^s_{(\beta\gamma)\alpha}(k).$$

For the anharmonic energy, we have a similar representation,

$$E_n = E_n^0 + \sum_\alpha \frac{1}{\alpha} \sum_{(p\nu)\alpha} p \sum_{(j)p+2} a_j \Pi^j_{(\beta\gamma)\nu}(0).$$

Expanding here the polynomials in terms of quantum numbers, we obtain the intriguing formula

$$\langle n|f|n+k\rangle = \sqrt{g_{n,n+k}} \sum_i \Phi_k^i \left(n + \frac{k+1}{2}\right)^i$$

for the one-dimensional case and

$$\langle n|f|n+k\rangle = \sqrt{g_{n_1,n_1+k_1}\cdots g_{n_r,n_r+k_r}} \sum_{i_1 \ldots i_r} \Phi^{i_1 \ldots i_r}_{k_1 \ldots k_r} \left(n_1 + \frac{k_1+1}{2}\right)^{i_1} \cdots \left(n_r + \frac{k_r+1}{2}\right)^{i_r}$$

for the many-dimensional case;

$$\Phi_k^i \text{ and } \Phi_{k_1\ldots k_r}^{i_1\ldots i_r}$$

are coefficients. The validity of this expansion follows from the condition of the symmetry of the matrix element of function f, namely,

$$(n|f|n+k) = (n+k|f|n).$$

The substitution of number k here by $-k$ yields

$$(n|f|n-k) = (n-k|f|n) = \sqrt{g_{n-k,n}} \sum_i \Phi_{-k}^i \left(n + \frac{1-k}{2}\right)^i.$$

We obtain this expansion alternatively with the aid of a formal substitution of n by $n - k$; that is, again,

$$(n|f|n-k) = (n-k|f|n) = \sqrt{g_{n-k,n}} \sum_i \Phi_k^i \left(n + \frac{1-k}{2}\right)^i.$$

Assuming $\Phi_{-k}^i = \Phi_k^i$, we ascertain that the expansion of $(n|f|n+k)$ in powers of $n + k/2 + 1/2$ is valid. Quantity $k/2$ ensures the symmetry of the matrix element, whereas factor $1/2$ appears because of the commutation relation between destruction operator η and creation operator η^+, $[\eta, \eta^+] = 1$. For the many-dimensional case, the validity of the expansion above is established in a similar manner.

The derived expansions in terms of quantum numbers hold for the matrix elements of an arbitrary physical function that is represented as an expansion in a power series in terms of creation and destruction operators. This consequence of perturbation theory calculations is trivial. The values of energy E_n are also expressible from the formula for $(n|f|n+k)$ in which $f = H$. Assuming $k = 0$, we obtain

$$E_n = \sum_{i_1 i_2 \ldots i_r} \Omega_{i_1 i_2 \ldots i_r} \left(n_1 + \frac{1}{2}\right)^{i_1} \left(n_2 + \frac{1}{2}\right)^{i_2} \ldots \left(n_r + \frac{1}{2}\right)^{i_r},$$

in which mechanical anharmonicity parameters $\Omega_{i_1 i_2 \ldots i_r}$ might be expressed through a_j and ω_k. To generalize our theory, we assume that quantity E_n is a function of quantum numbers $n_1 + 1/2, n_2 + 1/2, \ldots, n_r + 1/2$. Together with this dependence on the quantum numbers, energy E_n depends parametrically on some coefficients that govern the extent of anharmonicity. Appropriately choosing these coefficients and an explicit form of *quantum* function E_n, one might obtain a pertinent representation for anharmonicity.

We can heuristically determine a function Φ for the matrix element of a particular physical quantity $f(\xi)$, for instance, the dipolar moment, as a dependence on quantum number $n + k/2 + 1/2$:

$$\langle n|f|n+k\rangle = \sqrt{g_{n,n+k}}\Phi_k\left(n + \frac{k+1}{2}\right).$$

Functions Φ_k are arbitrarily expressible, for example,

$$\Phi_k = \theta_k(n+k/2+1/2)e^{-\phi_k(n+k/2+1/2)}, \quad \Phi_k = \theta_k(n+k/2+1/2)^{-1}, \text{ etc.},$$

with parameters θ_k and ϕ_k determined from experiments. In the present formalism, one might also phenomenologically construct a function $\Phi_{k_1 k_2 \ldots k_r}$ for a system with r variables. However, from the solution of the Schrödinger equation according to perturbation theory follows not the functions themselves but rather their expansions in terms of quantum numbers with coefficients Φ_k^i that characterize the exact influence of anharmonicity. These coefficients have no dependence on quantum numbers and have the dimension of initial function f. The introduction of the functions of quantum numbers is essentially a conversion to an "anharmonicity representation," which transcends the solution according to the perturbation theory. The study of these functions of quantum numbers with pertinent laws represents a special interest in physics today.

Other anharmonic models

Applying the method of factorization, we solve two useful problems for the eigenvalues of anharmonic Hamiltonians that describe, in an alternative manner, simple vibrational systems.[5] So, let

$$H = \frac{p_r^2}{2m} + V_r$$

be the general Hamiltonian of some physical system, which is a particle that moves in a given anharmonic potential V_r; $V_r = V(x)$. Here, r is the current coordinate of a particle of momentum p_r and mass m;

$$x = \frac{r - r_0}{r_0}$$

is the relative shift of coordinate r from its equilibrium value r_0. The scheme to determine the eigenvalues is simple. First, we postulate that

$$F = 2mH$$

Theory of anharmonicity

and, according to insight, we choose variable η_n. Then, on comparing two expressions for F_1, we find f_1. Value f_1 corresponds to the state of the system with the least energy E_0; $f_1 = 2mE_0$. Other quantities $f_2, f_3, \ldots$ follow from a comparison of two expressions for F_{n+1}. As

$$F|n\rangle = f_{n+1}|n\rangle,$$

the sought eigenvalues E_n, corresponding to eigenstates $|n\rangle$, are expressible through the formula

$$E_n = \frac{f_{n+1}}{2m}.$$

This scenario to find a solution is general; notice that our interest is focused on the energy levels of bound states and we ignore a continuous spectrum of energy.

Morse potential

As a first problem, we consider the widely known Morse's oscillator and show that the theory, which is developed in Chapter 1, generates the correct values for the energy levels of such an oscillator. We write the Hamiltonian in the form

$$H = \frac{p_r^2}{2m} + D(1 - e^{-a_M x})^2;$$

we assume

$$F = p_r^2 + A - 2Ae^{-a_M x} + Ae^{-2a_M x}$$

and

$$\eta_n = p_r + i(b_n + c_n e^{-a_M x}).$$

Here, D and a_M are the parameters of Morse's potential, $A = 2mD$; b_n and c_n are the real quantities, the explicit forms of which are to be defined.

We calculate $\eta_n^+ \eta_n$ as

$$\eta_n^+ \eta_n = p_r^2 - ic_n[e^{-a_M x}, p_r] + (b_n + c_n e^{-a_M x})^2$$
$$= p_r^2 + (2b_n - \hbar a_M / r_0) c_n e^{-a_M x} + b_n^2 + c_n^2 e^{-2a_M x},$$

in which we take into account that

$$[e^{-a_M x}, p_r] = i\hbar \frac{\partial}{\partial r} e^{-a_M x} = -i\hbar \frac{a_M}{r_0} e^{-a_M x}.$$

In an analogous manner, one finds

$$\eta_n \eta_n^+ = p_r^2 + (2b_n + \hbar a_M/r_0)c_n e^{-a_M x} + b_n^2 + c_n^2 e^{-2a_M x}.$$

Let us consider operator F_1:

$$F_1 = \eta_1^+ \eta_1 + f_1 = p_r^2 + (2b_1 - \hbar a_M/r_0)c_1 e^{-a_M x} + b_1^2 + c_1^2 e^{-2a_M x} + f_1.$$

From the other side,

$$F_1 = p_r^2 + A - 2A e^{-a_M x} + A e^{-2a_M x}.$$

On comparison, we obtain the following equations:

$$b_1 = \frac{\hbar a_M}{2r_0} - \frac{A}{c_1}, \quad c_1^2 = A \quad \text{and} \quad b_1^2 + f_1 = A.$$

Here, one might have two possible solutions. If $c_1 = -\sqrt{A}$, then

$$b_1 = \frac{\hbar a_M}{2r_0} + \sqrt{A} \quad \text{and} \quad f_1 = A - \left(\frac{\hbar a_M}{2r_0} + \sqrt{A}\right)^2,$$

whereas for $c_1 = \sqrt{A}$,

$$b_1 = \frac{\hbar a_M}{2r_0} - \sqrt{A} \quad \text{and} \quad f_1 = A - \left(\frac{\hbar a_M}{2r_0} - \sqrt{A}\right)^2.$$

Because, in the latter case, the value of quantity f_1 is greater, one chooses that second solution:

$$f_1 = A - \left(\frac{\hbar a_M}{2r_0} - \sqrt{A}\right)^2 = \frac{\hbar a_M}{r_0}\sqrt{A} - \left(\frac{\hbar a_M}{2r_0}\right)^2.$$

As $\hbar\omega = 2D a_M^2 \lambda^2$, in which ω is the vibrational frequency, and $\lambda = (1/r_0)\cdot\sqrt{(\hbar/m\omega)}$, then

$$a_M = \frac{1}{\lambda}\sqrt{\frac{\hbar\omega}{2D}} = r_0\omega\sqrt{\frac{m}{2D}}.$$

Consequently,

$$f_1 = m\hbar\omega - m\frac{(\hbar\omega)^2}{8D},$$

and the least eigenvalue of Hamiltonian H equals

$$E_0 = \frac{f_1}{2m} = \hbar\omega\left(0 + \frac{1}{2}\right) - \frac{(\hbar\omega)^2}{4D}\left(0 + \frac{1}{2}\right)^2.$$

We find other eigenvalues from a comparison of the two expressions for F_{n+1}. By definition,

$$\eta_n \eta_n^+ + f_n = \eta_{n+1}^+ \eta_{n+1} + f_{n+1};$$

that is,

$$p_r^2 + (2b_n + \hbar a_M/r_0)c_n e^{-a_M x} + b_n^2 + c_n^2 e^{-2a_M x} + f_n$$
$$= p_r^2 + (2b_{n+1} - \hbar a_M/r_0)c_{n+1} e^{-a_M x} + b_{n+1}^2 + c_{n+1}^2 e^{-2a_M x} + f_{n+1}.$$

Therefore,

$$c_{n+1}^2 = c_n^2,$$
$$(2b_{n+1} - \hbar a_M/r_0)c_{n+1} = (2b_n + \hbar a_M/r_0)c_n,$$

and

$$b_{n+1}^2 + f_{n+1} = b_n^2 + f_n.$$

We see that

$$c_{n+1}^2 = c_n^2 = c_{n-1}^2 = \cdots = c_1^2 = A;$$

discarding the solution $c_n = -\sqrt{A}$, one obtains

$$c_n = \sqrt{A}.$$

Furthermore,

$$b_{n+1} = b_n + \frac{\hbar a_M}{r_0} = b_{n-1} + 2\frac{\hbar a_M}{r_0} = \cdots = b_1 + n\frac{\hbar a_M}{r_0} = -\sqrt{A} + \left(n + \frac{1}{2}\right)\frac{\hbar a_M}{r_0}.$$

Finally,

$$b_{n+1}^2 + f_{n+1} = b_n^2 + f_n = \cdots = b_1^2 + f_1 = A$$

hence

$$f_{n+1} = A - b_{n+1}^2 = 2\frac{\hbar a_M}{r_0}\sqrt{A}\left(n+\frac{1}{2}\right) - (\hbar a_M/r_0)^2\left(n+\frac{1}{2}\right)^2.$$

Taking into account that

$$2\frac{\hbar a_M}{r_0}\sqrt{A} = 2m\hbar\omega \quad \text{and} \quad (\hbar a_M/r_0)^2 = 2m\frac{(\hbar\omega)^2}{4D},$$

we have

$$f_{n+1} = 2m\hbar\omega\left(n+\frac{1}{2}\right) - 2m\frac{(\hbar\omega)^2}{4D}\left(n+\frac{1}{2}\right)^2.$$

As $E_n = f_{n+1}/2m$, then

$$E_n = \hbar\omega\left(n+\frac{1}{2}\right) - \frac{(\hbar\omega)^2}{4D}\left(n+\frac{1}{2}\right)^2,$$

which is the required solution.

Generalized Morse problem

We proceed to complicate the Morse problem in considering the potential,

$$V(x) = D\left(\frac{1-e^{-ax}}{1-ke^{-ax}}\right)^2 = D\left(1 + 2\frac{k-1}{e^{ax}-k} + \frac{(k-1)^2}{(e^{ax}-k)^2}\right),$$

in which $a = (1-k)a_M$ and $|k| < 1$[11]; a_M, D, and k are adjustable parameters. The corresponding Hamiltonian has the form

$$H = \frac{p_r^2}{2m} + D\left(1 + 2\frac{k-1}{e^{ax}-k} + \frac{(k-1)^2}{(e^{ax}-k)^2}\right).$$

We define quantity F as $F = 2m(H - D)$, that is,

$$F = p_r^2 + \frac{A}{y} + \frac{B}{y^2},$$

in which $y = e^{ax} - k$, $A = 4mD(k-1)$, and $B = 2mD(k-1)^2$. Moreover, we use

$$\eta_n = p_r + i\left(b_n + \frac{c_n}{e^{ax}-k}\right);$$

Theory of anharmonicity

with $k = 0$, this variable transforms into analogous quantity η_n for the case of Morse's oscillator. We must define coefficients b_n and c_n.

According to our scenario, we begin from the calculation of $\eta_n^+ \eta_n$:

$$\eta_n^+ \eta_n = p_r^2 - ic_n[1/y, p_r] + b_n^2 + \frac{2b_n c_n}{y} + \frac{c_n^2}{y^2};$$

as

$$[1/y, p_r] = -i\hbar \frac{a}{r_0}\left(\frac{1}{y} + \frac{k}{y^2}\right),$$

then

$$\eta_n^+ \eta_n = p_r^2 + b_n^2 + \frac{1}{y}\left(2b_n c_n - \frac{\hbar a}{r_0} c_n\right) + \frac{1}{y^2}\left(c_n^2 - \frac{k\hbar a}{r_0} c_n\right);$$

analogously, we find that

$$\eta_n \eta_n^+ = p_r^2 + b_n^2 + \frac{1}{y}\left(2b_n c_n + \frac{\hbar a}{r_0} c_n\right) + \frac{1}{y^2}\left(c_n^2 + \frac{k\hbar a}{r_0} c_n\right).$$

Let us consider quantity F_1:

$$F_1 = \eta_1^+ \eta_1 + f_1 = p_r^2 + b_1^2 + \frac{1}{y}\left(2b_1 c_1 - \frac{\hbar a}{r_0} c_1\right) + \frac{1}{y^2}\left(c_1^2 - \frac{k\hbar a}{r_0} c_1\right) + f_1$$

$$\equiv p_r^2 + \frac{A}{y} + \frac{B}{y^2},$$

therefore

$$b_1^2 + f_1 = 0, \quad 2b_1 c_1 - \frac{\hbar a}{r_0} c_1 = A \quad \text{and} \quad c_1^2 - \frac{k\hbar a}{r_0} c_1 = B.$$

On solving the latter equation with respect to c_1, we obtain

$$c_1 = \frac{k\hbar a}{2r_0} \pm \sqrt{B + \left(\frac{k\hbar a}{2r_0}\right)^2};$$

consequently,

$$b_1 = \frac{A}{(k\hbar a/r_0) \pm 2\sqrt{B + (k\hbar a/2r_0)^2}} + \frac{\hbar a}{2r_0}.$$

Choosing b_1, which leads to the maximum value for f_1, we have

$$f_1 = -b_1^2 = -\left(\frac{\hbar a}{2r_0} + A\left(\frac{k\hbar a}{r_0} + 2\sqrt{B + \left(\frac{k\hbar a}{2r_0}\right)^2}\right)^{-1}\right)^2.$$

To calculate other quantities f_n, we consider the identity

$$\eta_n \eta_n^+ + f_n = \eta_{n+1}^+ \eta_{n+1} + f_{n+1}$$

or, in an explicit form,

$$p_r^2 + b_n^2 + \frac{1}{y}\left(2b_n c_n + \frac{\hbar a}{r_0} c_n\right) + \frac{1}{y^2}\left(c_n^2 + \frac{k\hbar a}{r_0} c_n\right) + f_n$$
$$= p_r^2 + b_{n+1}^2 + \frac{1}{y}\left(2b_{n+1} c_{n+1} - \frac{\hbar a}{r_0} c_{n+1}\right) + \frac{1}{y^2}\left(c_{n+1}^2 - \frac{k\hbar a}{r_0} c_{n+1}\right) + f_{n+1}.$$

On comparing the left and right parts of this identity, we find the equations

$$b_{n+1}^2 + f_{n+1} = b_n^2 + f_n = \cdots = b_1^2 + f_1 = 0,$$

$$2b_{n+1} c_{n+1} - \frac{\hbar a}{r_0} c_{n+1} = 2b_n c_n + \frac{\hbar a}{r_0} c_n, \text{ and } c_{n+1}\left(c_{n+1} - \frac{k\hbar a}{r_0}\right) = c_n\left(c_n + \frac{k\hbar a}{r_0}\right).$$

From the latter relation, discarding the solution $c_{n+1} = -c_n$, we obtain

$$c_{n+1} = c_n + \frac{k\hbar a}{r_0} = \cdots = c_1 + n\frac{k\hbar a}{r_0}.$$

In turn, we determine b_{n+1}, having summed all equations of this system

$$2b_{n+1} c_{n+1} - \gamma c_{n+1} = 2b_n c_n + \gamma c_n,$$
$$2b_n c_n - \gamma c_n = 2b_{n-1} c_{n-1} + \gamma c_{n-1},$$
$$\vdots$$
$$2b_2 c_2 - \gamma c_2 = 2b_1 c_1 + \gamma c_1,$$

in which $\gamma = \hbar a / r_0$. As a result,

$$2b_{n+1} c_{n+1} = 2b_1 c_1 + \gamma(c_2 + \cdots + c_n + c_{n+1}) + \gamma(c_1 + \cdots + c_{n-1} + c_n);$$

as

$$2b_1c_1 - \gamma c_1 = A$$

and

$$c_1 + \cdots + c_{n-1} + c_n = nc_1 + k\gamma \frac{n(n-1)}{2},$$

then

$$2b_{n+1}c_{n+1} = A + \gamma c_{n+1} + 2\gamma n\left(c_1 + k\gamma\frac{n-1}{2}\right),$$

therefore

$$b_{n+1} = \frac{A + 2\gamma n(c_1 + k\gamma(n-1)/2)}{2(c_1 + nk\gamma)} + \frac{\gamma}{2} = \frac{A - k\gamma^2 n(n+1)}{2(c_1 + nk\gamma)} + \gamma\left(n + \frac{1}{2}\right).$$

Supposing

$$n' = n + \frac{1}{2}, n(n+1) = n'^2 - \frac{1}{4} \text{ and } c_1 + nk\gamma = k\gamma n' + \sqrt{B + \left(\frac{k\gamma}{2}\right)^2},$$

we simplify the obtained expression for b_{n+1}. We have

$$b_{n+1} = \frac{A - k\gamma^2 n'^2 + (k\gamma^2/4)}{2k\gamma\left(n' + \text{sgn}(k)\sqrt{(B/(k\gamma)^2) + (1/4)}\right)} + \gamma n'$$

$$\equiv \frac{(A/k\gamma) + (\gamma/4) - \gamma((n'+Q) - Q)^2}{2(n'+Q)} + \gamma n',$$

in which $Q = \text{sgn}(k)\sqrt{(B/(k\gamma)^2) + (1/4)}$; hence,

$$b_{n+1} = \frac{(A/k\gamma) + (\gamma/4) - \gamma Q^2}{2(n'+Q)} + \frac{\gamma}{2}(n'+Q).$$

Finally, taking into account that

$$\frac{A}{k\gamma} + \frac{\gamma}{4} - \gamma Q^2 = \frac{2mD}{\gamma}\left(1 - \frac{1}{k^2}\right),$$

we obtain

$$f_{n+1} = -b_{n+1}^2 = -mD(1-k^{-2}) - \frac{m^2 D^2}{\gamma^2} \cdot \frac{(1-k^{-2})^2}{(n'+Q)^2} - \frac{\gamma^2}{4}(n'+Q)^2.$$

The eigenvalues of Hamiltonian H are thus equal to

$$\frac{f_{n+1}}{2m} + D = \frac{D}{2}(1+k^{-2}) - \frac{mD^2}{2\gamma^2} \cdot \frac{(1-k^{-2})^2}{(n'+Q)^2} - \frac{\gamma^2}{8m}(n'+Q)^2$$

and

$$E_n = \frac{D}{2}\left[1 + k^{-2} - \frac{mDr_0^2}{\hbar^2 a^2} \cdot \frac{(1-k^{-2})^2}{(n+1/2+Q)^2} - \frac{\hbar^2 a^2}{4mDr_0^2}(n+1/2+Q)^2\right]$$

are the sought energy levels of the generalized Morse's oscillator;

$$Q = \operatorname{sgn}(k)\sqrt{\frac{2mDr_0^2}{\hbar^2 a^2}(1-k^{-1})^2 + \frac{1}{4}}.$$

Quantum fields

4

Creation and destruction operators

Until this point, we have considered a description in terms of the coordinates and momenta; both latter quantities, according to the general ideology of quantum theory, are expressible through quantized values, whereas the fields, concerned with these variables, remain classical. A correct description requires a revision of the theory. One might achieve this purpose with the aid of second quantization. The occupation numbers of separate particles in particular states become variables after second or repeated quantization. The new formalism constitutes a basis of quantum electrodynamics; with its aid, one might solve the problems of quantization of not only the electromagnetic field but also the one-particle fields of the Schrödinger, Klein–Fock–Gordon, and Dirac equations.

We begin with the consideration of a boson field.[12] Let field operators $\varphi(\mathbf{r})$ and $\varphi^+(\mathbf{r})$ satisfy the commutation relations

$$[\varphi(\mathbf{r}), \varphi^+(\mathbf{r}')] = \delta(\mathbf{r} - \mathbf{r}'),$$

$$[\varphi(\mathbf{r}), \varphi(\mathbf{r}')] = 0, \text{ and } [\varphi^+(\mathbf{r}), \varphi^+(\mathbf{r}')] = 0,$$

in which $\mathbf{r}$ and $\mathbf{r}'$ are the radius vectors of two arbitrary points; all quantities are here taken at one, and the same, moment of time. Through these terms, the one-particle function of Hamilton for the particle moving in a field with potential $V(\mathbf{r})$ has the form

$$H = \int \varphi^+ \left(-\frac{\hbar^2}{2m}\nabla^2 + V(\mathbf{r})\right)\varphi \, d\tau,$$

in which m is the mass of the particle and $d\tau$ is an element of volume. One might arrive at this expression through traditional formalism of analytical mechanics of fields with Lagrange's function density

$$\mathcal{L} = i\hbar\varphi^+\dot{\varphi} - \frac{\hbar^2}{2m}(\nabla\varphi^+)(\nabla\varphi) - V\varphi^+\varphi.$$

However, we prefer to postulate that a Hamiltonian, not a Lagrangian, is appropriate here.

Let us consider an equation of the motion for $\varphi(\mathbf{r})$; we have

$$i\hbar\dot\varphi = [\varphi, H] = \left[\varphi(\mathbf{r}), \int \varphi^+(\mathbf{r}')\left(-\frac{\hbar^2}{2m}\nabla'^2\right)\varphi(\mathbf{r}')d\tau'\right] + \left[\varphi(\mathbf{r}), \int \varphi^+(\mathbf{r}')V(\mathbf{r}')\varphi(\mathbf{r}')d\tau'\right]$$

$$= \int [\varphi(\mathbf{r}), \varphi^+(\mathbf{r}')]\left(-\frac{\hbar^2}{2m}\nabla'^2 + V(\mathbf{r}')\right)\varphi(\mathbf{r}')d\tau' = -\frac{\hbar^2}{2m}\nabla^2\varphi(\mathbf{r}) + V(\mathbf{r})\varphi(\mathbf{r}).$$

This result represents the equation of Schrödinger. In an analogous manner, we obtain for the case of variable $\varphi^+(\mathbf{r})$:

$$-i\hbar\dot\varphi^+ = -\frac{\hbar^2}{2m}\nabla^2\varphi^+(\mathbf{r}) + V(\mathbf{r})\varphi^+(\mathbf{r}).$$

One must bear in mind that φ and φ^+ are operators already.

If some complete orthonormal system of wavefunctions $\psi_i(\mathbf{r})$ exists, then one might determine the field operators at an arbitrary moment of time t through the expansions

$$\varphi(\mathbf{r}, t) = \sum_i a_i(t)\psi_i(\mathbf{r}) \text{ and } \varphi^+(\mathbf{r}, t) = \sum_i \psi_i^*(\mathbf{r})a_i^+(t),$$

in which a_i and a_i^+ are the familiar operators of destruction and creation. Through the orthonormality of functions ψ_i, we have

$$a_i(t) = \int \varphi(\mathbf{r}, t)\psi_i^*(\mathbf{r})d\tau \text{ and } a_i^+(t) = \int \psi_i(\mathbf{r})\varphi^+(\mathbf{r}, t)d\tau;$$

consequently,

$$[a_i, a_i^+] = \int \psi_i^* \psi_i'[\varphi(\mathbf{r}, t), \varphi^+(\mathbf{r}', t)]d\tau\, d\tau' = \int |\psi_i|^2\, d\tau = 1,$$

$$[a_i, a_j] = 0, \text{ and } [a_i^+, a_j^+] = 0;$$

as $[a_i, a_j^+] = 0$ at $i \neq j$,

$$[a_i, a_j^+] = \delta_{ij}.$$

Quantities a_i and a_i^+ operate in an abstract space of occupation numbers n_i,

$$a_i|n_i\rangle = \sqrt{n_i}|n_i - 1\rangle, \text{ and } a_i^+|n_i\rangle = \sqrt{n_i + 1}|n_i + 1\rangle,$$

in which $|n_i\rangle$ are the eigenvectors of quantity $N_i = a_i^+ a_i$;

$$N_i|n_i\rangle = n_i|n_i\rangle.$$

The operator N for total particle number is defined through the expression

$$N = \int \varphi^+ \varphi d\tau = \sum_{i,j} a_i^+ a_j \int \psi_i^* \psi_j d\tau = \sum_i a_i^+ a_i = \sum_i N_i.$$

For the vacuum state, in which there is no particle, for all values of i, this identity is satisfied:

$$N_i|0\rangle = 0.$$

On acting on the vector of the vacuum state with creation operators on successive occasions of sufficient number, one might obtain the state with an arbitrary number of particles. In a general case,

$$|n_1, n_2, \ldots\rangle = \left\{ \prod_i \frac{1}{\sqrt{n_i!}} (a_i^+)^{n_i} \right\} |0\rangle$$

is the vector describing n_1 particles in state 1, n_2 particles in state 2, and so on. Notice that operator a_i^+ creates a particle in a state with wavefunction $\psi_i(\mathbf{r})$, which incarnates the first quantization. The transfer into Hilbert space of the occupation numbers corresponds to the second quantization.

If $\psi_i(\mathbf{r})$ is the eigenfunction belonging to eigenvalue ε_i of this chosen Hamiltonian, then

$$H = \sum_{i,j} a_i^+ a_j \int \psi_i^* \left(-\frac{\hbar^2}{2m} \nabla^2 + V \right) \psi_j d\tau \to \sum_i n_i \varepsilon_i.$$

There is no difficulty in understanding the latter result: it is the energy representation for the field of bosons; n_1 particles occupy the state with energy ε_1, n_2 particles occupy the state with energy ε_2, and so on. All quantities N_i, together with the total number of particles, N, in this case, become the constants of the motion.

In terms of the field operators, one might define the quantity total momentum of the field as

$$\mathbf{P} = \int \varphi^+ \mathbf{p} \varphi d\tau = \sum_{i,j} a_i^+ a_j \int \psi_i^* \mathbf{p} \psi_j d\tau.$$

Let functions ψ_i of the first quantization be the eigenfunctions of the momentum operator $\mathbf{p}$; we have

$$\mathbf{P} \to \sum_i n_i \mathbf{p}_i,$$

in which $\mathbf{p}_i$ are the eigenvalues of momentum corresponding to functions ψ_i. This heuristic expression in the representation of the occupation numbers indicates that n_1 particles possess momentum $\mathbf{p}_1$, n_2 particles possess momentum $\mathbf{p}_2$, and so on. For free particles, $V = 0$ and, in this case, eigenfunctions ψ_i are the same for both Hamiltonian and momentum.

These aspects of second quantization for a boson field are principal. If Bose particles of various types figure in a problem, for each type one should introduce its own operators of creation and destruction. The operators belonging to various boson fields commute with each other. Notice that all drawn conclusions correspond to one and the same moment of time t. To evaluate the temporal variation of field quantities, one must apply Heisenberg's equations of motion.

We proceed to a fermion field. As for the case of bosons, we introduce field operators in a form of the expansions,

$$\varphi(\mathbf{r},t) = \sum_i c_i(t)\psi_i(\mathbf{r}) \text{ and } \varphi^+(\mathbf{r},t) = \sum_i \psi_i^*(\mathbf{r})c_i^+(t).$$

Quantities c_i and c_i^+ are operators, and $\psi_1(\mathbf{r})$, $\psi_2(\mathbf{r})$, ... represent orthonormal wavefunctions in a complete set. For fermions, as is known, one must satisfy Pauli's exclusion principle; therefore, we refrain from applying the typical commutation relations but, as Jordan and Wigner showed, we might apply the following anticommutation relations:

$$[\varphi(\mathbf{r},t), \varphi^+(\mathbf{r}',t)]_+ = \varphi(\mathbf{r},t)\varphi^+(\mathbf{r}',t) + \varphi^+(\mathbf{r}',t)\varphi(\mathbf{r},t) = \delta(\mathbf{r}-\mathbf{r}'),$$

$$[\varphi(\mathbf{r},t), \varphi(\mathbf{r}',t)]_+ = 0, \text{ and } [\varphi^+(\mathbf{r},t), \varphi^+(\mathbf{r}',t)]_+ = 0;$$

$$[c_i, c_j^+]_+ = \delta_{ij}, [c_i, c_j]_+ = [c_i^+, c_j^+]_+ = 0.$$

To elucidate the meaning of the latter expressions, we introduce two possible state vectors for fermions

$$|1\rangle = \begin{pmatrix} 1 \\ 0 \end{pmatrix} \text{ and } |0\rangle = \begin{pmatrix} 0 \\ 1 \end{pmatrix}.$$

In the former case, a fermion occupies a state; in the latter case, a state is unoccupied. Operators c and c^+, in this case, are expressible through the 2×2 Jordan–Wigner matrices

$$c = \begin{pmatrix} 0 & 0 \\ 1 & 0 \end{pmatrix} = \frac{1}{2}(\sigma_x - i\sigma_y) \text{ and } c^+ = \begin{pmatrix} 0 & 1 \\ 0 & 0 \end{pmatrix} = \frac{1}{2}(\sigma_x + i\sigma_y),$$

in which σ_x and σ_y are 2×2 Pauli matrices. One sees that

$$cc^+ + c^+c = \begin{pmatrix} 1 & 0 \\ 0 & 1 \end{pmatrix}, [c,c]_+ = [c^+,c^+]_+ = \begin{pmatrix} 0 & 0 \\ 0 & 0 \end{pmatrix}.$$

In a manner analogous to that of the field of bosons, we define the particle number operator $N = c^+ c$:

$$c^+ c|1\rangle = 1 \cdot |1\rangle \text{ and } c^+ c|0\rangle = 0 \cdot |0\rangle;$$

$$N^2 = c^+ c c^+ c = c^+(1 - c^+ c)c = c^+ c = N.$$

Eigenvalues n of the particle number operator equal 1 and 0, as required through Pauli's principle. Operator c destroys and operator c^+ creates a fermion in a given state, i.e.,

$$c|1\rangle = |0\rangle \text{ and } c^+|0\rangle = |1\rangle.$$

It is important that

$$c|0\rangle = 0 \text{ and } c^+|1\rangle = 0.$$

Furthermore, we might construct a space of occupation numbers in which the field operators act. For this purpose, one must act with creation operators on the vacuum state. We have

$$|0, 0, \ldots, 1_i, \ldots, 0\rangle = c_i^+ |0, 0, \ldots, 0_i, \ldots, 0\rangle,$$
$$|0, 0, \ldots, 1_i, \ldots, 1_j, \ldots, 0\rangle = c_i^+ c_j^+ |0, 0, \ldots, 0_i, \ldots, 0_j, \ldots, 0\rangle, \text{ and so on.}$$

Is the location of various operators c and c^+ before the vacuum state vector important? The answer is affirmative. To understand this fact, we consider a two-particle state with vector $|1, 1\rangle$ and initially act on it with operator c_1. As a result,

$$c_1|1, 1\rangle = c_1 c_1^+ c_2^+ |0, 0\rangle = (1 - c_1^+ c_1) c_2^+ |0, 0\rangle = |0, 1\rangle.$$

We then act on $|1, 1\rangle$ with operator c_2; we have

$$c_2|1, 1\rangle = c_2 c_1^+ c_2^+ |0, 0\rangle = -c_1^+(1 - c_2^+ c_2)|0, 0\rangle = -|1, 0\rangle.$$

As we see through the fact that operators c and c^+ fail to commute — they anticommute — in the latter expression a minus sign appears. Thus, if for a particular order there exists occupied state i, which is located on the left from state j, on the vector on which we act with either operator c_j or operator c_j^+, the minus sign arises; in an opposite case, the plus sign remains. This rule is expressible through simple relations,

$$c_j|n_j\rangle = \vartheta_j n_j |1 - n_j\rangle \text{ and } c_j^+|n_j\rangle = \vartheta_j(1 - n_j)|1 - n_j\rangle,$$

in which

$$\vartheta_j = (-1)^{n_1 + n_2 + \cdots + n_{j-1}}$$

characterizes the number of occupied states on the left from j. To determine the order of location for various fermion field operators before the vacuum state vector, one must write the action of quantities c^+ and c in the form of a normal product in which all c^+ appear on the left from c.

In conclusion, we consider a heuristic expression for a Hamiltonian in terms of field Fermi operators for which we choose the solutions of Schrödinger's one-particle equation, corresponding to eigenvalues ε_i, as wavefunctions $\psi_i(\mathbf{r})$. We have

$$H = \int \varphi^+ \left(-\frac{\hbar^2}{2m}\nabla^2 + V\right)\varphi d\tau = \sum_{i,j} c_i^+ c_j \int \psi_i^* \left(-\frac{\hbar^2}{2m}\nabla^2 + V\right)\psi_j d\tau = \sum_i \varepsilon_i c_i^+ c_i.$$

If $V = 0$, then $\varepsilon_i = \mathbf{p}_i^2/2m$; analogously to the conclusions for bosons, the total momentum of the field of fermions is, in this case, given by the formula

$$\mathbf{P} = \sum_i \mathbf{p}_i c_i^+ c_i.$$

Free scalar field

To construct quantum field theory, we begin with a description of a particle with spin zero. For a *field* φ in this case, Lagrange's function density has a form

$$\mathcal{L} = \frac{\varphi_{,\mu}\varphi^{,\mu}}{2} - \frac{m_0^2 c^2 \varphi^2}{2\hbar^2},$$

and the equation of the motion leads to the Klein–Fock–Gordon equation:

$$\frac{\partial \mathcal{L}}{\partial \varphi} - \frac{\partial}{\partial x^\mu}\frac{\partial \mathcal{L}}{\partial \varphi_{,\mu}} = -\left(\frac{m_0^2 c^2}{\hbar^2} + \frac{\partial^2}{\partial x_\mu \partial x^\mu}\right)\varphi = 0;$$

here, m_0 denotes the mass of the particle and $x^\mu = (ct, \mathbf{r})$ are the space–time coordinates (see section 'Dirac's equation' in Chapter 2). A canonically conjugate momentum equals

$$v = \frac{\partial \mathcal{L}}{\partial \dot\varphi} = \frac{\dot\varphi}{c^2},$$

in which $\dot\varphi = c\varphi^0$. In what follows, we assume that $c = 1$, and only if necessary, for instance, as in the case when one considers a fermion field of electrons and positrons, do we indicate constant c explicitly. In the final expressions, one might easily restore the speed of light from considerations of the dimensions.

At a fixed moment of time, the classical Poisson brackets have the form

$$\{\varphi(x_0, \mathbf{r}), v(x_0, \mathbf{r}')\} = \delta(\mathbf{r} - \mathbf{r}')$$

and

$$\{\varphi(x_0,\mathbf{r}),\varphi(x_0,\mathbf{r}')\} = \{v(x_0,\mathbf{r}),v(x_0,\mathbf{r}')\} = 0.$$

Our scenario is the same as in the case of the nonrelativistic Schrödinger's equation. We initially work with a classical field, then replace Poisson brackets with commutators, and transfer to the quantum theory.

Hence, we consider a real scalar field corresponding to a particle with zero rest mass.[13] For this purpose, we assume $m_0 = 0$ and redefine Lagrange's function density, having multiplied it by $1/4\pi$, then

$$\mathcal{L} = \frac{\varphi_{,\mu}\varphi^{,\mu}}{8\pi}$$

and, consequently,

$$\varphi_{,\mu}{}^{,\mu} = 0.$$

The new momentum decreases and becomes equal to $\dot{\varphi}/4\pi$. In the theory, however, we maintain the preceding momentum

$$v = \dot{\varphi}.$$

The Poisson brackets herewith are slightly altered:

$$\{\varphi(\mathbf{r}),v(\mathbf{r}')\} = 4\pi\delta(\mathbf{r}-\mathbf{r}')$$

and

$$\{\varphi(\mathbf{r}),\varphi(\mathbf{r}')\} = \{v(\mathbf{r}),v(\mathbf{r}')\} = 0.$$

Here, we omit argument x_0 because we imply a definite moment of time. Hereafter we use such designations, with x_0 equal to some concrete value.

We determine Hamilton's function density,

$$\mathcal{H} = \frac{\dot{\varphi}}{4\pi}\cdot\dot{\varphi} - \mathcal{L} = \frac{1}{8\pi}(v^2 - \varphi_{,i}\varphi^{,i}),$$

and a Hamiltonian accordingly,

$$H = \int \mathcal{H}d\mathbf{r} = \frac{1}{8\pi}\int(v^2 - \varphi_{,i}\varphi^{,i})d\mathbf{r}.$$

The equations of motion repeat the results that are already known,

$$\dot{\varphi}(\mathbf{r}) = \{\varphi(\mathbf{r}),H\} = \frac{1}{8\pi}\int\{\varphi(\mathbf{r}),v^2(\mathbf{r}')\}d\mathbf{r}' = v(\mathbf{r})$$

and

$$\dot{v}(\mathbf{r}) = \{v(\mathbf{r}), H\} = -\frac{1}{8\pi}\int\{v(\mathbf{r}), \varphi_{,i}\varphi^{i}\}d\mathbf{r}' = -\varphi^{i}{}_{,i}(\mathbf{r});$$

to obtain the second equation, we take into account that

$$\{v(\mathbf{r}), \varphi^{i}(\mathbf{r}')\} = -4\pi\frac{\partial\delta(\mathbf{r}'-\mathbf{r})}{\partial x'_i}.$$

We proceed to other variables, expanding our field into a Fourier integral in plane waves,

$$\varphi(x_0, \mathbf{r}) = \int (\varphi_k^c e^{ik_\mu x^\mu} + \overline{\varphi}_k^c e^{-ik_\mu x^\mu})d\mathbf{k},$$

in which

$$k_\mu x^\mu = k_0 x_0 - \mathbf{k}\cdot\mathbf{r} \text{ and } k_0 = \omega_k = |\mathbf{k}| > 0.$$

Fixing time and assuming

$$\varphi_k = \varphi_k^c e^{ik_0 x_0},$$

we rewrite our expansion in the form

$$\varphi(\mathbf{r}) = \int(\varphi_k e^{-i\mathbf{k}\cdot\mathbf{r}} + \overline{\varphi}_k e^{i\mathbf{k}\cdot\mathbf{r}})d\mathbf{k} = \int(\varphi_k + \overline{\varphi}_{-k})e^{-i\mathbf{k}\cdot\mathbf{r}}d\mathbf{k}.$$

The inverse transformation yields

$$\varphi_k + \overline{\varphi}_{-k} = (2\pi)^{-3}\int\varphi(\mathbf{r})e^{i\mathbf{k}\cdot\mathbf{r}}d\mathbf{r}.$$

Differentiating $\varphi(\mathbf{r})$ with respect to x_0, we determine the expansion for conjugate momentum,

$$v(\mathbf{r}) = \frac{\partial\varphi(\mathbf{r})}{\partial x_0} = i\int|\mathbf{k}|(\varphi_k - \overline{\varphi}_{-k})e^{-i\mathbf{k}\cdot\mathbf{r}}d\mathbf{k};$$

we also invert it, resulting in

$$i|\mathbf{k}|(\varphi_k - \overline{\varphi}_{-k}) = (2\pi)^{-3}\int v(\mathbf{r})e^{i\mathbf{k}\cdot\mathbf{r}}d\mathbf{r}.$$

As already clear, the new variables are now φ_k and $\overline{\varphi}_k$. This moment of general theory is highly important because it is generally accepted to operate by quantities φ_k^c and $\overline{\varphi}_k^c$ rather than by φ_k and $\overline{\varphi}_k$. Quantities φ_k^c and $\overline{\varphi}_k^c$ are the constant coefficients of the expansion into a Fourier integral — they are independent of time. In the literature, $\overline{\varphi}_k^c$ corresponds to destruction operator a_k, and φ_k^c corresponds to creation operator a_k^+. From the one side, it is natural; from the other side, considering the interaction of the scalar field with other fields, we fail to correctly determine the coefficients independent of time, φ_k^c and $\overline{\varphi}_k^c$. Dirac first noticed this distinction. Following him, we adhere to this point of view, that is, our variables are φ_k and $\overline{\varphi}_k$.

Proceeding from the expressions for the inverse Fourier images of quantities $\varphi(\mathbf{r})$ and $\upsilon(\mathbf{r})$, we calculate the Poisson brackets for new variables:

$$i|\mathbf{k}'|\{\varphi_k + \overline{\varphi}_{-k}, \varphi_{k'} - \overline{\varphi}_{-k'}\} = (2\pi)^{-6} \int\int \{\varphi(\mathbf{r}), \upsilon(\mathbf{r}')\} e^{i\mathbf{k}\cdot\mathbf{r}} e^{i\mathbf{k}'\cdot\mathbf{r}'} d\mathbf{r}\, d\mathbf{r}'$$

$$= 4\pi \cdot (2\pi)^{-6} \int e^{i(\mathbf{k}+\mathbf{k}')\cdot\mathbf{r}} d\mathbf{r} = (2\pi^2)^{-1} \delta(\mathbf{k}+\mathbf{k}').$$

Furthermore, as

$$\{\varphi(\mathbf{r}), \varphi(\mathbf{r}')\} = \{\upsilon(\mathbf{r}), \upsilon(\mathbf{r}')\} = 0,$$

then

$$\{\varphi_k + \overline{\varphi}_{-k}, \varphi_{k'} + \overline{\varphi}_{-k'}\} = 0 \text{ and } \{\varphi_k - \overline{\varphi}_{-k}, \varphi_{k'} - \overline{\varphi}_{-k'}\} = 0.$$

On combining these Poisson brackets with bracket

$$\{\varphi_k + \overline{\varphi}_{-k}, \varphi_{k'} - \overline{\varphi}_{-k'}\}$$

that was found here, we obtain

$$\{\varphi_k + \overline{\varphi}_{-k}, \varphi_{k'}\} = -\frac{i}{4\pi^2|\mathbf{k}'|} \delta(\mathbf{k}+\mathbf{k}')$$

and

$$\{\varphi_k, \varphi_{k'} - \overline{\varphi}_{-k'}\} = -\frac{i}{4\pi^2|\mathbf{k}'|} \delta(\mathbf{k}+\mathbf{k}') = -\frac{i}{4\pi^2|\mathbf{k}|} \delta(\mathbf{k}+\mathbf{k}').$$

Substituting k with k', we rewrite the former expression in the form

$$\{\varphi_k, \varphi_{k'} + \overline{\varphi}_{-k'}\} = \frac{i}{4\pi^2|\mathbf{k}|} \delta(\mathbf{k}+\mathbf{k}').$$

In comparison, we find the sought Poisson brackets for the new variables:

$$\{\varphi_k, \overline{\varphi}_{-k'}\} = \frac{i}{4\pi^2|\mathbf{k}|}\delta(\mathbf{k}+\mathbf{k}') \text{ or } \{\varphi_k, \overline{\varphi}_{k'}\} = \frac{i}{4\pi^2|\mathbf{k}|}\delta(\mathbf{k}-\mathbf{k}'), \{\varphi_k, \varphi_{k'}\} = 0.$$

One should transform the Hamiltonian. We have

$$\int v^2 d\mathbf{r} = -\iiint |\mathbf{k}||\mathbf{k}'|(\varphi_k - \overline{\varphi}_{-k})(\varphi_{k'} - \overline{\varphi}_{-k'})e^{-i(\mathbf{k}+\mathbf{k}')\cdot\mathbf{r}}d\mathbf{k}\,d\mathbf{k}'\,d\mathbf{r}$$

$$= -(2\pi)^3 \iint |\mathbf{k}||\mathbf{k}'|(\varphi_k - \overline{\varphi}_{-k})(\varphi_{k'} - \overline{\varphi}_{-k'})\delta(\mathbf{k}+\mathbf{k}')d\mathbf{k}\,d\mathbf{k}'$$

$$= -(2\pi)^3 \int |\mathbf{k}|^2(\varphi_k - \overline{\varphi}_{-k})(\varphi_{-k} - \overline{\varphi}_k)d\mathbf{k}$$

and, furthermore, implying the summation with respect to i ($i = 1, 2$ and 3),

$$\int (\varphi^i)^2 d\mathbf{r} = -\iiint k_i(\varphi_k + \overline{\varphi}_{-k})e^{-i\mathbf{k}\cdot\mathbf{r}} k'_i(\varphi_{k'} + \overline{\varphi}_{-k'})e^{-i\mathbf{k}'\cdot\mathbf{r}}d\mathbf{k}\,d\mathbf{k}'\,d\mathbf{r}$$

$$= -(2\pi)^3 \iint k_i k'_i(\varphi_k + \overline{\varphi}_{-k})(\varphi_{k'} + \overline{\varphi}_{-k'})\delta(\mathbf{k}+\mathbf{k}')d\mathbf{k}\,d\mathbf{k}'$$

$$= (2\pi)^3 \int |\mathbf{k}|^2(\varphi_k + \overline{\varphi}_{-k})(\varphi_{-k} + \overline{\varphi}_k)d\mathbf{k};$$

consequently,

$$H = 2\pi^2 \int |\mathbf{k}|^2(\varphi_k \overline{\varphi}_k + \varphi_{-k}\overline{\varphi}_{-k})d\mathbf{k} = 4\pi^2 \int |\mathbf{k}|^2 \varphi_k \overline{\varphi}_k d\mathbf{k}.$$

We proceed to quantum theory. For this purpose, we replace the Poisson brackets with the commutators, having multiplied them by $i\hbar$; as a result,

$$[\varphi_k, \varphi_{k'}] = 0$$

and

$$[\varphi_k, \overline{\varphi}_{k'}] = -\frac{\hbar}{4\pi^2|\mathbf{k}|}\delta(\mathbf{k}-\mathbf{k}').$$

According to the Fock representation for Bose–Einstein statistics, one might redefine the current variables; namely, if we use

$$\varphi_k = \frac{1}{2\pi}\sqrt{\frac{\hbar}{|\mathbf{k}|}} \cdot \iota_k^+ \text{ and } \overline{\varphi}_k = \frac{1}{2\pi}\sqrt{\frac{\hbar}{|\mathbf{k}|}} \cdot \iota_k,$$

then

$$[\iota_k, \iota_{k'}^+] = \delta(\mathbf{k} - \mathbf{k}').$$

In terms of the Fock operators, the Hamiltonian has the form

$$H = \frac{\hbar}{2} \int |\mathbf{k}| \left(\iota_k^+ \iota_k + \iota_k \iota_k^+ \right) d\mathbf{k} = \hbar \int |\mathbf{k}| \iota_k^+ \iota_k d\mathbf{k} + \text{infinite constant},$$

which substantially corresponds to the harmonic oscillators in a set. One might generally neglect this infinite number. It corresponds to the total energy $\sum \hbar \omega_k / 2$ of the zero-point vibrations of oscillators of infinite number. Such a discarding of the infinite constant is correct. Proceeding to quantum considerations from classical, we can directly postulate the Hamiltonian in the form $\hbar \int |\mathbf{k}| \iota_k^+ \iota_k d\mathbf{k}$. Not the physical interest of the energies but rather their differences, i.e., the frequencies, are obviously independent of the constant in addition to the Hamiltonian.

In conclusion, we discuss the relativistic invariance of the developed theory. We have identified the Poisson brackets with the commutators and thus have quantized the classically built theory. However, the quantities in commutators have belonged to one determinate moment of time. To prove the relativistic invariance, one must calculate the same commutator but for two separate space–time points, i.e., commutator

$$[\varphi(x_0, \mathbf{r}), \varphi(x_0', \mathbf{r}')],$$

and one must define whether the result is Lorentz invariant.

So, taking into account that

$$\left[\varphi_k^c, \overline{\varphi}_{k'}^c \right] = \left[\varphi_k, \overline{\varphi}_{k'} \right] e^{-i(k_0 - k_0')x_0} = -\frac{\hbar}{4\pi^2 |\mathbf{k}|} \delta(\mathbf{k} - \mathbf{k}'),$$

in which $\varphi_k = \varphi_k^c e^{ik_0 x_0}$ and $k_0 = |\mathbf{k}|$, we have

$$[\varphi(x_0, \mathbf{r}), \varphi(x_0', \mathbf{r}')] = \iint \left[\varphi_k^c e^{ik_\mu x^\mu} + \overline{\varphi}_k^c e^{-ik_\mu x^\mu}, \varphi_{k'}^c e^{ik_\mu' x'^\mu} + \overline{\varphi}_{k'}^c e^{-ik_\mu' x'^\mu} \right] d\mathbf{k}\, d\mathbf{k}'$$

$$= -\frac{\hbar}{4\pi^2} \int \frac{1}{|\mathbf{k}|} (e^{ik_\mu(x^\mu - x'^\mu)} - e^{-ik_\mu(x^\mu - x'^\mu)}) d\mathbf{k}$$

$$= -\frac{\hbar}{4\pi^2} \iint \frac{1}{|\mathbf{k}|} (e^{ik_\mu(x^\mu - x'^\mu)} - e^{-ik_\mu(x^\mu - x'^\mu)}) \delta(k_0 - |\mathbf{k}|) d\mathbf{k}\, dk_0$$

$$= -\frac{\hbar}{4\pi^2} \int \frac{1}{|\mathbf{k}|} e^{ik_\mu(x^\mu - x'^\mu)} (\delta(k_0 - |\mathbf{k}|) - \delta(k_0 + |\mathbf{k}|)) d^4k.$$

To prove the relativistic invariance of the obtained quantity, we introduce an auxiliary function

$$\Delta(b) = \frac{2b_0}{|b_0|}\delta(b_\mu b^\mu),$$

which plays an important role in field theory. This function is obviously Lorentz invariant; $b_0/|b_0|$ gives only the plus or minus sign. We subsequently identify four-vector b_μ with k_μ, whereas $k_0 > 0$, so our choice is correct. The simple transformation of a Δ-function yields

$$\Delta(b) = \frac{2b_0}{|b_0|}\delta(b_0^2 - \mathbf{b}^2) = \frac{2b_0}{|b_0|}\delta((b_0 - |\mathbf{b}|)(b_0 + |\mathbf{b}|)) = \frac{2b_0}{|b_0|}\left(\frac{\delta(b_0 - |\mathbf{b}|)}{|b_0 + |\mathbf{b}||} + \frac{\delta(b_0 + |\mathbf{b}|)}{|b_0 - |\mathbf{b}||}\right);$$

hence,

$$\Delta(b) = \frac{1}{|\mathbf{b}|}(\delta(b_0 - |\mathbf{b}|) - \delta(b_0 + |\mathbf{b}|)).$$

Assuming properly that $b_\mu = k_\mu$, we return to the commutator,

$$[\varphi(x_0, \mathbf{r}), \varphi(x_0', \mathbf{r}')] = -\frac{\hbar}{4\pi^2}\int e^{ik_\mu(x^\mu - x'^\mu)}\Delta(k)d^4k.$$

Our hypothesis is confirmed and the result of the calculations turns out to be Lorentz invariant. Moreover,

$$\int e^{ik_\mu x^\mu}\Delta(k)d^4k = 4\pi^2 i\Delta(x),$$

such that eventually

$$[\varphi(x_0, \mathbf{r}), \varphi(x_0', \mathbf{r}')] = -i\hbar\Delta(x - x').$$

Quantization of electromagnetic field

We consider an electromagnetic field in the absence of external charges. The scenario is as before — for a basis we take the classical canonical theory with some modification of a physical nature and we then develop quantum formalism, as first performed by Fermi.[14] However, the method of canonical quantization is not unique. The quantization of an electromagnetic field is associated with some difficulties. According to classical electrodynamics, a field is generally described with the aid of four-vector A_μ of potential, whereas the physical observables are quantities

$$F_{\mu\nu} = A_{\nu,\mu} - A_{\mu,\nu}$$

that represent the electric field and magnetic field vectors. Operating with quantities A_μ, we preserve, in a natural manner, the relativistic invariance but we encounter a problem because the components of a four-vector of potential fail to be independent dynamical variables. In addition to the superfluous variables, there exists another problem — determining a rigorously positive Hamilton's function density. Let us surmount these difficulties of the quantization of electromagnetic field.

Lagrange's function density, which yields Maxwell's equations, has the form

$$\mathcal{L} = -\frac{1}{16\pi} F_{\mu\nu} F^{\mu\nu} - \frac{1}{8\pi} A_\mu{}^{,\mu} A_\nu{}^{,\nu}.$$

If we additionally take into account a Lorenz gauge $A_\mu{}^{,\mu} = 0$, $\mathcal{L}$ becomes simplified. We postpone this condition and transform $\mathcal{L}$:

$$F_{\mu\nu} F^{\mu\nu} = (A_{\nu,\mu} - A_{\mu,\nu})(A^{\nu,\mu} - A^{\mu,\nu}) = 2A_{\mu,\nu}(A^{\mu,\nu} - A^{\nu,\mu}),$$

$$A_\mu{}^{,\mu} A_\nu{}^{,\nu} = \left(A_\mu A_\nu{}^{,\nu}\right)^{,\mu} - A_\mu A_\nu{}^{,\nu\mu} = \left[(A_\mu A_\nu{}^{,\nu})^{,\mu} - (A_\mu A_\nu{}^{,\mu})^{,\nu}\right] + A_\mu{}^{,\nu} A_\nu{}^{,\mu}.$$

Because the expression in square brackets is given by the total derivative, it does not affect the Euler–Lagrange equations because the variation of the Lagrangian for this part is exactly equal to zero. Excluding this expression, we write Lagrange's function density in the form

$$\mathcal{L} = -\frac{1}{8\pi} A_{\mu,\nu} A^{\mu,\nu},$$

such that $\mathcal{L}$ explicitly resembles Lagrange's function density of a scalar field that we considered in the preceding section; however, we have here four components of the four-vector of potential rather than one function. The equations of motion yield the wave equations

$$A_{\mu,\nu}{}^{,\nu} = 0$$

for each component, respectively.

One might introduce canonical variables, considering A_μ as independent scalar fields. On A_μ, one must then impose the additional conditions

$$A_\mu{}^{,\mu} = 0,$$

which are necessary to eliminate the superfluous dynamical degrees of freedom. Fermi used such restrictions, but he realized them in the sense of weak constraints

$$A_\mu{}^{,\mu} \approx 0.$$

Accepting this, we should bear in mind that, proceeding to the corresponding operators from the classical variables, one must perform the condition

$$A_\mu{}^{,\mu}|P\rangle = 0,$$

not the condition of strong constraint $A_\mu{}^{,\mu} = 0$; $|P\rangle$ is an arbitrary vector.

Fermi's ideology

For each of the four fields A_μ, we apply the formalism developed in the preceding section, which allows one to introduce the canonical variables. Let

$$v_\mu(\mathbf{r}) = \frac{\partial A_\mu(\mathbf{r})}{\partial x_0}$$

be the momentum conjugate with variable A_μ. For the Poisson brackets, we therefore have the following relations:

$$\{A_i(\mathbf{r}), v_i(\mathbf{r}')\} = 4\pi\delta(\mathbf{r} - \mathbf{r}')$$

and

$$\{A_0(\mathbf{r}), v_0(\mathbf{r}')\} = -4\pi\delta(\mathbf{r} - \mathbf{r}');$$

other brackets are equal to zero. In the relation for the time-like components, the minus sign appears because the *true* momentum $\partial \mathcal{L}/\partial A_0{}^{,0}$ turns out to be negative; it equals

$$-A_0{}^{,0}/4\pi.$$

In the theory we use the momenta, which are 4π-times the truly conjugate momenta.

With the aid of the flat space metric, we combine the relations for the brackets and represent them in the form

$$\{A_\mu(\mathbf{r}), v_\nu(\mathbf{r}')\} = -4\pi g_{\mu\nu}\delta(\mathbf{r} - \mathbf{r}'),$$

in which $g^{\mu\nu} = 0$ for $\mu \neq \nu$, $g^{00} = 1$, and $g^{ii} = -1$; $A_0 = A^0$ and $A_i = -A^i$. We represent the Hamiltonian as the sum of the four Hamiltonians for the free scalar fields, which are associated with A_μ; as a result,

$$H = -\frac{1}{8\pi}\int (v_\mu(\mathbf{r})v^\mu(\mathbf{r}) - A_{\mu,i}(\mathbf{r})A^{\mu,i}(\mathbf{r}))d\mathbf{r}.$$

Quantum fields

Furthermore, we expand the field into a Fourier integral,

$$A_\mu(\mathbf{r}) = \int (A_{\mu k} e^{i\mathbf{k}\cdot\mathbf{r}} + \overline{A}_{\mu k} e^{-i\mathbf{k}\cdot\mathbf{r}}) d\mathbf{k},$$

in which $A_{\mu k} = A_{\mu k}^c e^{ik_0 x_0}$, and convert the Hamiltonian to the form

$$H = -4\pi^2 \int |\mathbf{k}|^2 A_{\mu k} \overline{A}_k^\mu d\mathbf{k} = 4\pi^2 \int |\mathbf{k}|^2 (A_{1k}\overline{A}_{1k} + A_{2k}\overline{A}_{2k} + A_{3k}\overline{A}_{3k} - A_{0k}\overline{A}_{0k}) d\mathbf{k}.$$

Hamilton's function density includes a purely negative term

$$-4\pi^2 |\mathbf{k}|^2 A_{0k} \overline{A}_{0k}$$

that is undesirable. For this reason, many authors prefer a so-called radiation gauge, with

$$A_0 = 0.$$

In this case, the Lorentz invariance and gradient covariance of the theory are entirely lost but, as we see later, there is nothing wrong with the negative term contributing to the energy of fields.

Let us write the expressions

$$\{A_{\mu k}, \overline{A}_{\nu k'}\} = -\frac{ig_{\mu\nu}}{4\pi^2 |\mathbf{k}|} \delta(\mathbf{k} - \mathbf{k}')$$

and

$$\{A_{\mu k}, A_{\nu k'}\} = \{\overline{A}_{\mu k}, \overline{A}_{\nu k'}\} = 0$$

for the Poisson brackets on the corresponding Fourier components; we proceed to quantum theory. Multiplying the right part of the Poisson brackets by $i\hbar$, we obtain the commutation relations

$$[A_{\mu k}, \overline{A}_{\nu k'}] = \frac{\hbar g_{\mu\nu}}{4\pi^2 |\mathbf{k}|} \delta(\mathbf{k} - \mathbf{k}')$$

and

$$[A_{\mu k}, A_{\nu k'}] = [\overline{A}_{\mu k}, \overline{A}_{\nu k'}] = 0.$$

We postulate the Hamiltonian

$$H = -4\pi^2 \int |\mathbf{k}|^2 A_{\mu k} \overline{A}_k^\mu d\mathbf{k}$$

without changes, having excluded only the infinite constant that is concerned with the energy of the vacuum. We do not duplicate the results of classical theory; quantities $A_{\mu k}$ are now operators, and the Poisson brackets are replaced with the commutators. The classical theory serves as a particular initial point at which to correctly choose Hamilton's variables in accordance with both the relativistic invariance and the general principles of analytical mechanics of fields.

Regarding the subsidiary conditions for quantities $A_{\mu}{}^{,\mu}$, these quantities must commute with all variables that have a physical meaning. Such variables are readily understood to be components $F_{\mu\nu}$ of the electric field and magnetic field vectors, which must commute with each other. The physical meaning is that the chosen conditions are satisfied in principle and are compatible by themselves; they are eventually required as additional relations connecting the dynamical variables.

Following Fermi, we demand the performance of our conditions in a weak sense

$$A_{\mu}{}^{,\mu} \approx 0,$$

such that

$$A_{\mu}{}^{,\mu}|P\rangle = 0$$

for arbitrary vector $|P\rangle$. The action of a true physical variable on $|P\rangle$ must also yield a physical vector, for instance, vector $|P'\rangle$; otherwise, we treat with a nonphysical quantity. In our case,

$$F_{\mu\nu}|P\rangle = |P'\rangle.$$

In addition, one must require the performance of the condition

$$A_{\sigma}{}^{,\sigma}|P'\rangle = 0, \quad \text{i.e.,} \quad A_{\sigma}{}^{,\sigma}F_{\mu\nu}|P\rangle = 0;$$

hence,

$$[F_{\mu\nu}, A_{\sigma}{}^{,\sigma}]|P\rangle = 0,$$

because $F_{\mu\nu}A_{\sigma}{}^{,\sigma}|P\rangle = 0$. In an analogous manner, we find the condition of compatibility,

$$[A_{\mu}{}^{,\mu}, A_{\nu}{}^{,\nu}]|P\rangle = 0.$$

The conditions are enunciated; we show that they are satisfied through a direct calculation of the commutators.

For two separate space–time points,

$$[F_{\mu\nu}, A_{\sigma}{}^{,\sigma}] = [A_{\nu,\mu}(x_0, \mathbf{r}) - A_{\mu,\nu}(x_0, \mathbf{r}), A_{\sigma}{}^{,\sigma}(x'_0, \mathbf{r}')]$$
$$= \frac{\partial^2}{\partial x^{\mu}\partial x'_{\sigma}}[A_{\nu}(x_0, \mathbf{r}), A_{\sigma}(x'_0, \mathbf{r}')] - \frac{\partial^2}{\partial x^{\nu}\partial x'_{\sigma}}[A_{\mu}(x_0, \mathbf{r}), A_{\sigma}(x'_0, \mathbf{r}')].$$

Recalling that

$$[A_\mu(x_0, \mathbf{r}), A_\nu(x'_0, \mathbf{r}')] = i\hbar g_{\mu\nu}\Delta(x - x'),$$

because each component A_μ here is identical with a separate scalar field φ (see the preceding section), we have

$$\left[F_{\mu\nu}, A_\sigma{}^{,\sigma}\right] = i\hbar \frac{\partial^2 \Delta(x - x')}{\partial x^\mu \partial x'^\nu} - i\hbar \frac{\partial^2 \Delta(x - x')}{\partial x^\nu \partial x'^\mu}.$$

Through an expansion of a function $\Delta(x - x')$ into a Fourier integral,

$$\frac{\partial^2 \Delta(x - x')}{\partial x^\mu \partial x'^\nu} = \frac{\partial^2}{\partial x^\mu \partial x'^\nu} \frac{1}{4\pi^2 i} \int e^{ik_\mu(x^\mu - x'^\mu)} \Delta(k) d^4 k = -\frac{\partial^2 \Delta(x - x')}{\partial x^\mu \partial x^\nu},$$

consequently,

$$[F_{\mu\nu}, A_\sigma{}^{,\sigma}] = 0.$$

We calculate analogously the second commutator,

$$\left[A_\mu{}^{,\mu}, A_\nu{}^{,\nu}\right] = \frac{\partial^2}{\partial x_\mu \partial x'_\nu} \left[A_\mu(x_0, \mathbf{r}), A_\nu(x'_0, \mathbf{r}')\right] = i\hbar \frac{\partial^2}{\partial x_\mu \partial x'^\mu} \Delta(x - x') = 0,$$

in which, for a transition to a Fourier integral for $\Delta(x - x')$, we take into account the identity $k_\mu k^\mu = 0$. We have thus achieved more than was required. Our commutators are equal to zero; the stated conditions are consequently satisfied in a strong, not weak, sense. The theory becomes self-consistent, despite that, as natural variables, we have nonphysical quantities A_μ.

When we apply a Fourier expansion for A_μ, the conditions of constraint

$$\frac{\partial A_\mu(x_0, \mathbf{r})}{\partial x_\mu} = -i \int k^\mu \left(A^c_{\mu k} e^{ik^\nu x_\nu} - \overline{A}^c_{\mu k} e^{-ik^\nu x_\nu}\right) d\mathbf{k} \approx 0$$

for the spatial components become transformed into the relations

$$k^\mu A^c_{\mu k} \approx 0 \text{ and } k^\mu \overline{A}^c_{\mu k} \approx 0$$

or

$$k^\mu A_{\mu k} \approx 0 \text{ and } k^\mu \overline{A}_{\mu k} \approx 0,$$

which mutually connect the corresponding Fourier components of the four-potential of electromagnetic field. The relations obtained are simple and allow the possibility of solving both the problem of superfluous variables and the question regarding the negative part of energy. To show this, we separate our field into the longitudinal and transverse parts. For A_3 associated with the longitudinal part, two other spatial components A_1 and A_2 correspond to the transverse waves.

Excluding from consideration transverse components A_1 and A_2, we write the Hamiltonian

$$H = 4\pi^2 \int |\mathbf{k}|^2 (A_{3k}\overline{A}_{3k} - A_{0k}\overline{A}_{0k})d\mathbf{k}.$$

Discarding the infinite constant, we have

$$H = 4\pi^2 \int |\mathbf{k}|^2 (A_{3k}\overline{A}_{3k} - \overline{A}_{0k}A_{0k})d\mathbf{k}.$$

We also write the conditions of constraint

$$A_{0k} - A_{3k} \approx 0 \text{ and } \overline{A}_{0k} - \overline{A}_{3k} \approx 0,$$

in which we use

$$k_1 = k_2 = 0$$

for the transverse part and

$$k_3 = k_0$$

for the longitudinal part. We multiply the first equation of constraint by $\overline{A}_{0k} + \overline{A}_{3k}$ and we multiply the second by $A_{0k} + A_{3k}$; we add these relations to obtain

$$A_{0k}\overline{A}_{0k} + \overline{A}_{0k}A_{0k} - (A_{3k}\overline{A}_{3k} + \overline{A}_{3k}A_{3k}) \approx 0;$$

therefore,

$$2(\overline{A}_{0k}A_{0k} - A_{3k}\overline{A}_{3k}) + [A_{0k}, \overline{A}_{0k}] + [A_{3k}, \overline{A}_{3k}] \approx 0.$$

Here, both commutators are substantially ordinary numbers, although they are infinite; because they have opposite signs, they compensate each other. Our conditions become converted into a weak equality,

$$\overline{A}_{0k}A_{0k} - A_{3k}\overline{A}_{3k} \approx 0.$$

Our Hamiltonian, acting on any physical vector, contains no negative term because the longitudinal spatial part and the negative time-like part are exactly compensated. In what follows, we consider only physical vectors; the problem regarding the negative part of the energy thereby ceases to be of concern.

Having solved the problems of superfluous variables and of a negative contribution in the Hamiltonian density, we have put the theory at a disadvantage; namely, we have deprived it of relativistic invariance because a separation of fields into longitudinal and transverse parts is not preserved under a Lorentz transformation. We must sacrifice something to advance.

Electron–positron Dirac field

We have considered an electromagnetic field and have quantized it according to Bose–Einstein statistics. Let us now study an important fermion field — the field of electrons and positrons. Proceeding as before, we initially construct the classical theory, which yields the correct equations, and then we proceed to the quantum theory through the replacement of the classical Poisson brackets by the anticommutation relations in well-known accordance with Fermi–Dirac statistics.

We write Lagrange's function density as

$$\mathcal{L} = c\overline{\psi}(\gamma^\mu p_\mu - mc)\psi = c\overline{\psi}\left(i\hbar\gamma^\mu \frac{\partial}{\partial x^\mu} - mc\right)\psi;$$

here, we apply the designations introduced in Chapter 2. In this case, the equation of motion yields the Dirac equation,

$$\frac{\partial \mathcal{L}}{\partial \overline{\psi}} = c(\gamma^\mu p_\mu - mc)\psi = 0,$$

and a canonically conjugate momentum equals

$$v = \frac{\partial \mathcal{L}}{\partial \dot{\psi}} = \frac{1}{c}\frac{\partial \mathcal{L}}{\partial \psi_{,0}} = \overline{\psi} \cdot i\hbar\gamma^0 = i\hbar\psi^+.$$

Note that ψ is the spinor with four components ψ_μ, which are represented as a column matrix; ψ^+ has components ψ^*_μ and the form of a row matrix. Consequently, function $\mathcal{L}$ represents a function of eight independent variables. In turn, $\overline{\psi} = \psi^+\gamma^0$ signifies the Dirac adjoint spinor. If possible, we omit the index that indicates a component of a spinor, but one must always bear this in mind.

We define Hamilton's function density,

$$\mathcal{H} = v\dot{\psi} - \mathcal{L} = vc\psi_{,0} - i\hbar c\overline{\psi}\gamma^\mu\psi_{,\mu} + mc^2\overline{\psi}\psi = -i\hbar c\psi^+\alpha^i\psi_{,i} + c\psi^+\beta\psi,$$

in which $\gamma^\mu = (\gamma^0, \gamma^0\alpha)$ and $\beta = mc\gamma^0$. Hence, the Hamiltonian is

$$H = c\int(-i\hbar\psi^+\alpha^i\psi_{,i} + \psi^+\beta\psi)d\mathbf{r}.$$

Furthermore, for the conjugate variables at a fixed moment of time, the conditions are performed in the form of Poisson brackets,

$$\{\psi_\mu(\mathbf{r}), \psi_\nu(\mathbf{r}')\} = \{\psi_\mu^+(\mathbf{r}), \psi_\nu^+(\mathbf{r}')\} = 0$$

and

$$\{\psi_\mu(\mathbf{r}), v_\nu(\mathbf{r}')\} = \delta_{\mu\nu}\delta(\mathbf{r} - \mathbf{r}') \text{ or } \{\psi_\mu(\mathbf{r}), \psi_\nu^+(\mathbf{r}')\} = \frac{1}{i\hbar}\delta_{\mu\nu}\delta(\mathbf{r} - \mathbf{r}').$$

Using these conditions, one might easily show that Hamilton's equation of motion

$$\dot\psi = \{\psi, H\}$$

leads to the correct equation of a Dirac field,

$$i\hbar(\psi^0 + \alpha_i\psi^{i}) - \beta\psi = 0.$$

Spinor ψ^+ also satisfies the Dirac equation,

$$i\hbar(\psi^{+,0} + \psi^{+,i}\alpha_i) + \psi^+\beta = 0.$$

Following the algorithm, as in the case of a scalar field, we expand the field variables into a Fourier integral,

$$\psi(\mathbf{r}) = (2\pi\hbar)^{-3/2}\int \tau(\mathbf{p})e^{i\mathbf{r}\cdot\mathbf{p}/\hbar}\,d\mathbf{p}.$$

The inverse transformation obviously has a similar form,

$$\tau(\mathbf{p}) = (2\pi\hbar)^{-3/2}\int \psi(\mathbf{r})e^{-i\mathbf{r}\cdot\mathbf{p}/\hbar}\,d\mathbf{r},$$

in which $\tau(\mathbf{p})$ consists of the four components that represent a column matrix. Furthermore,

$$\psi_{,i}(\mathbf{r}) = -\frac{i}{\hbar(2\pi\hbar)^{3/2}}\int p_i\tau(\mathbf{p})e^{i\mathbf{r}\cdot\mathbf{p}/\hbar}\,d\mathbf{p};$$

in terms of the Fourier components, consequently, the Hamiltonian equals

$$H = \frac{c}{(2\pi\hbar)^{3/2}}\iint \psi^+(\mathbf{r})(\boldsymbol{\alpha}\cdot\mathbf{p} + \beta)\tau(\mathbf{p})e^{i\mathbf{r}\cdot\mathbf{p}/\hbar}\,d\mathbf{p}\,d\mathbf{r}$$

$$= c\int \tau^+(\mathbf{p})(\boldsymbol{\alpha}\cdot\mathbf{p} + \beta)\tau(\mathbf{p})d\mathbf{p}.$$

To obtain more convenient form for use of the expression of the Hamiltonian, we perform a unitary transformation with the aid of matrix $\mathcal{U}(\mathbf{p})$,

$$\mathcal{U}(\mathbf{p})\tau(\mathbf{p}) = \varphi(\mathbf{p}) \text{ or explicitly } \mathcal{U}_{\mu\sigma}(\mathbf{p})\tau^\sigma(\mathbf{p}) = \varphi_\mu(\mathbf{p});$$

we return to the Hamiltonian,

$$H = c\int \tau^+(\mathbf{p})\overline{\mathcal{U}}(\mathbf{p})\mathcal{U}(\mathbf{p})(\alpha \cdot \mathbf{p} + \beta)\overline{\mathcal{U}}(\mathbf{p})\mathcal{U}(\mathbf{p})\tau(\mathbf{p})d\mathbf{p}$$
$$= c\int \varphi^+(\mathbf{p})[\mathcal{U}(\mathbf{p})(\alpha \cdot \mathbf{p} + \beta)\overline{\mathcal{U}}(\mathbf{p})]\varphi(\mathbf{p})d\mathbf{p},$$

and we take into account that $\mathcal{U}(\mathbf{p})\overline{\mathcal{U}}(\mathbf{p}) = I$. At this point it is clear that "more convenient form" should be understood as the diagonal form of matrix $\alpha \cdot \mathbf{p} + \beta$. Hence, transformation $\mathcal{U}(\mathbf{p})$ is chosen so that matrix $\mathcal{U}(\mathbf{p})(\alpha \cdot \mathbf{p} + \beta)\overline{\mathcal{U}}(\mathbf{p})$ comprises only diagonal non-zero elements. Moreover, we have no interest in knowing the explicit form of $\mathcal{U}(\mathbf{p})$; our interest is the final form of this matrix. We can obtain this form from simple heuristic considerations, namely that unitary transformation preserves the eigenvalues of the initial matrix, the square of which equals

$$(\alpha \cdot \mathbf{p} + \beta)^2 = (\mathbf{p}^2 + (mc)^2)I.$$

Thus, one might easily conclude that the eigenvalues of matrix $\alpha \cdot \mathbf{p} + \beta$ are the doubly degenerate values $\pm\sqrt{\mathbf{p}^2 + (mc)^2}$; hence,

$$\mathcal{U}(\mathbf{p})(\alpha \cdot \mathbf{p} + \beta)\overline{\mathcal{U}}(\mathbf{p}) = \sqrt{\mathbf{p}^2 + (mc)^2}\begin{pmatrix} I & 0 \\ 0 & -I \end{pmatrix},$$

in which I is a 2×2 unit matrix. The Hamiltonian thereby acquires a simple form,

$$H = c\int \sqrt{\mathbf{p}^2 + (mc)^2}\,\varphi^+(\mathbf{p})\begin{pmatrix} I & 0 \\ 0 & -I \end{pmatrix}\varphi(\mathbf{p})d\mathbf{p}$$
$$= c\int \sqrt{\mathbf{p}^2 + (mc)^2}\,\big(\varphi_1^*(\mathbf{p})\varphi_1(\mathbf{p}) + \varphi_2^*(\mathbf{p})\varphi_2(\mathbf{p}) - \varphi_3^*(\mathbf{p})\varphi_3(\mathbf{p}) - \varphi_4^*(\mathbf{p})\varphi_4(\mathbf{p})\big)d\mathbf{p}.$$

These are the classical conclusions. Is it possible to take them as a base of quantum theory? Let us reply to this question gradually. We begin with the expressions for Poisson brackets. To proceed to the quantum theory, one should replace the Poisson brackets with the anticommutators and multiply them by $i\hbar$; as a result,

$$\left[\psi_\mu(\mathbf{r}), \psi_\nu^+(\mathbf{r}')\right]_+ = \delta_{\mu\nu}\delta(\mathbf{r} - \mathbf{r}'),$$

$$\left[\psi_\mu(\mathbf{r}), \psi_\nu(\mathbf{r}')\right]_+ = \left[\psi_\mu^+(\mathbf{r}), \psi_\nu^+(\mathbf{r}')\right]_+ = 0.$$

However, our present variables are Fourier components $\varphi(\mathbf{p})$, not $\tau(\mathbf{p})$. Let us suppose that $\varphi_\mu(\mathbf{p})$ satisfy the relations

$$\left[\varphi_\mu(\mathbf{p}), \varphi_\nu^+(\mathbf{p}')\right]_+ = \delta_{\mu\nu}\delta(\mathbf{p} - \mathbf{p}')$$

and

$$\left[\varphi_\mu(\mathbf{p}), \varphi_\nu(\mathbf{p}')\right]_+ = \left[\varphi_\mu^+(\mathbf{p}), \varphi_\nu^+(\mathbf{p}')\right]_+ = 0,$$

which are guessed intuitively; we verify whether they are consistent with the anticommutators for $\psi(\mathbf{r})$.

So, as $\tau_\mu(\mathbf{p}) = \overline{\mathcal{U}}_{\mu\sigma}(\mathbf{p})\varphi^\sigma(\mathbf{p})$, then

$$\begin{aligned}\left[\tau_\mu(\mathbf{p}), \tau_\nu^+(\mathbf{p}')\right]_+ &= \overline{\mathcal{U}}_{\mu\sigma}(\mathbf{p})\mathcal{U}_{\nu\rho}(\mathbf{p}')[\varphi^\sigma(\mathbf{p}), \varphi^{+\rho}(\mathbf{p}')]_+ \\ &= \overline{\mathcal{U}}_{\mu\sigma}(\mathbf{p})\mathcal{U}_{\nu\rho}(\mathbf{p}')\delta^{\sigma\rho}\delta(\mathbf{p} - \mathbf{p}') = \delta_{\mu\nu}\delta(\mathbf{p} - \mathbf{p}'),\end{aligned}$$

and also

$$\left[\tau_\mu(\mathbf{p}), \tau_\nu(\mathbf{p}')\right]_+ = \left[\tau_\mu^+(\mathbf{p}), \tau_\nu^+(\mathbf{p}')\right]_+ = 0.$$

The unitary transformation thus has no influence on the anticommutation relations. Further, having applied the Fourier representation, we directly calculate $\left[\psi_\mu(\mathbf{r}), \psi_\nu^+(\mathbf{r}')\right]_+$:

$$\begin{aligned}\left[\psi_\mu(\mathbf{r}), \psi_\nu^+(\mathbf{r}')\right]_+ &= (2\pi\hbar)^{-3}\iint\left[\tau_\mu(\mathbf{p}), \tau_\nu^+(\mathbf{p}')\right]_+ e^{i\mathbf{r}\cdot\mathbf{p}/\hbar}e^{-i\mathbf{r}'\cdot\mathbf{p}'/\hbar}d\mathbf{p}\,d\mathbf{p}' \\ &= \delta_{\mu\nu}(2\pi\hbar)^{-3}\int e^{i\mathbf{p}\cdot(\mathbf{r}-\mathbf{r}')/\hbar}d\mathbf{p} = \delta_{\mu\nu}\delta(\mathbf{r} - \mathbf{r}'),\end{aligned}$$

and, of course,

$$\left[\psi_\mu(\mathbf{r}), \psi_\nu(\mathbf{r}')\right]_+ = \left[\psi_\mu^+(\mathbf{r}), \psi_\nu^+(\mathbf{r}')\right]_+ = 0.$$

We have confirmed the correctness of the suppositions regarding the quantum conditions in the form of anticommutation relations for the electron–positron Dirac field. Let us proceed to the choice of the Hamiltonian.

It turns out that, having replaced only functions φ and φ^* by field operators φ and φ^+, respectively, we cannot simply postulate the classical Hamiltonian in quantum theory. If we use

$$H = c\int\sqrt{\mathbf{p}^2 + (mc)^2}\left(\varphi_1^+(\mathbf{p})\varphi_1(\mathbf{p}) + \varphi_2^+(\mathbf{p})\varphi_2(\mathbf{p}) - \varphi_3^+(\mathbf{p})\varphi_3(\mathbf{p}) - \varphi_4^+(\mathbf{p})\varphi_4(\mathbf{p})\right)d\mathbf{p},$$

then, because of the two minuses before the third and fourth terms in parentheses, in addition to positive physical values of the energy, our Hamiltonian would also yield negative energies, which are impermissible. However, we proceed from this Hamiltonian, having preliminarily transformed it with the aid of the anticommutation relations as follows,

$$H = c \int \sqrt{\mathbf{p}^2 + (mc)^2} \left(\varphi_1^+(\mathbf{p})\varphi_1(\mathbf{p}) + \varphi_2^+(\mathbf{p})\varphi_2(\mathbf{p}) + \varphi_3(\mathbf{p})\varphi_3^+(\mathbf{p}) + \varphi_4(\mathbf{p})\varphi_4^+(\mathbf{p}) \right) d\mathbf{p} + C,$$

in which

$$C = -c \int \sqrt{\mathbf{p}^2 + (mc)^2} \left([\varphi_3(\mathbf{p}), \varphi_3^+(\mathbf{p})]_+ + [\varphi_4(\mathbf{p}), \varphi_4^+(\mathbf{p})]_+ \right) d\mathbf{p}.$$

Each commutator yields here $\delta(0)$; hence, quantity C is independent of the field variables, and is simply an infinite constant. Like the case of a scalar field, we discard the infinite c-number because it has no influence on physically observable results, which one might obtain with the aid of our Hamiltonian. Thus, eventually,

$$H = c \int \sqrt{\mathbf{p}^2 + (mc)^2} \left(\varphi_1^+(\mathbf{p})\varphi_1(\mathbf{p}) + \varphi_2^+(\mathbf{p})\varphi_2(\mathbf{p}) + \varphi_3(\mathbf{p})\varphi_3^+(\mathbf{p}) + \varphi_4(\mathbf{p})\varphi_4^+(\mathbf{p}) \right) d\mathbf{p}.$$

The constructed Hamiltonian describes the particles with exclusively positive values of energy. Therefore, we can define a vacuum state through a simple equation,

$$H|0\rangle = 0.$$

This equation yields a series of trivial relations,

$$\varphi_1(\mathbf{p})|0\rangle = \varphi_2(\mathbf{p})|0\rangle = \varphi_3^+(\mathbf{p})|0\rangle = \varphi_4^+(\mathbf{p})|0\rangle = 0,$$

which show that operators φ_1, φ_2, φ_3^+, and φ_4^+ are essentially destruction operators, whereas φ_1^+, φ_2^+, φ_3, and φ_4 are creation operators. Regarding the first couple of variables, we have total clarity; these variables correspond to electrons with a positive energy. According to the algorithm of second quantization, $\varphi_1^+(\mathbf{p})$ and $\varphi_2^+(\mathbf{p})$ create whereas $\varphi_1(\mathbf{p})$ and $\varphi_2(\mathbf{p})$ destroy an electron with momentum $\mathbf{p}$. A somewhat different situation arises for the second couple of variables, which, as is already clear, correspond to positrons. Here, the creation and destruction operators are interchanged, but there is nothing inappropriate. In the standard treatment with positive and negative values of energy, $\varphi_3^+(\mathbf{p})$ and $\varphi_4^+(\mathbf{p})$ correspond to creation whereas $\varphi_3(\mathbf{p})$ and $\varphi_4(\mathbf{p})$ correspond to destruction of an electron with negative energy and momentum $\mathbf{p}$. In our case, $\varphi_3(\mathbf{p})$ and $\varphi_4(\mathbf{p})$ play the role of creation operators for real particles with momentum $-\mathbf{p}$; the minus sign appears because, according to the common procedure of second quantization, we destroy an electron with momentum $\mathbf{p}$. We interpret these particles as positrons.

The theory of the Dirac field is substantially constructed. During our consideration, we merely have discarded the infinite constant that physically represents the energy of the "sea" of electrons, which possesses negative energy. However, infinity arises not only in the Hamiltonian. In particular, we consider an expression for the total charge of a system,

$$Q = -e\int\left(\varphi_1^+(\mathbf{p})\varphi_1(\mathbf{p}) + \varphi_2^+(\mathbf{p})\varphi_2(\mathbf{p})\right)d\mathbf{p} + e\int\left(\varphi_3(\mathbf{p})\varphi_3^+(\mathbf{p}) + \varphi_4(\mathbf{p})\varphi_4^+(\mathbf{p})\right)d\mathbf{p}$$

$$= -e\int\left(\varphi_1^+(\mathbf{p})\varphi_1(\mathbf{p}) + \varphi_2^+(\mathbf{p})\varphi_2(\mathbf{p}) + \varphi_3^+(\mathbf{p})\varphi_3(\mathbf{p}) + \varphi_4^+(\mathbf{p})\varphi_4(\mathbf{p})\right)d\mathbf{p} + (\sim \delta(0)).$$

Here, the infinite constant, which corresponds to the total charge of the sea of electrons with negative energy, also appears. Discarding it, we obtain

$$Q = -e\int \varphi^+(\mathbf{p})\varphi(\mathbf{p})d\mathbf{p}$$

or, returning from the Fourier components to variables $\psi(\mathbf{r})$,

$$Q = -e\int \psi^+(\mathbf{r})\psi(\mathbf{r})d\mathbf{r} \equiv \int \rho_e d\mathbf{r}.$$

In an analogous manner, one might interpret a current density,

$$\mathbf{j}_e = -ec\psi^+(\mathbf{r})\boldsymbol{\alpha}\psi(\mathbf{r}).$$

As a result, we have a theory with localized densities of energy, charge, and current.

Interaction picture

We have considered the procedures of quantization of electron–positron and electromagnetic fields. These fields were free and this case has no particular interest. We proceed to involve the interacting fields. In quantum electrodynamics, the interaction of fields leads to the appearance of an anomalous magnetic moment for an electron and to an electromagnetic shift of energy levels.

Let us introduce an interaction into the theory. So, we have the field of electrons and positrons with Hamiltonian

$$H_e = \int \left(-i\hbar\psi^+\alpha^i\psi_{,i} + \psi^+\beta\psi\right)d\mathbf{r},$$

in which the speed of light does not explicitly appear because we set $c = 1$. We have an electromagnetic field that is described by Hamiltonian

$$H_f = -\frac{1}{8\pi}\int \left(v_\mu v^\mu - A_{\mu,i}A^{\mu,i}\right)\mathrm{d}\mathbf{r},$$

plus the interaction with Hamilton's function H_{int}, which we take directly from classical electrodynamics as

$$H_{\text{int}} = \int A_\mu j^\mu \mathrm{d}\mathbf{r},$$

in which $j^\mu = (\rho_e, \mathbf{j}_e)$ is the four-vector of a current density. For convenience, we omit index e at j^μ. Let us clarify this choice. First, the energy of interaction of electronic and electromagnetic fields must simultaneously contain the variables of both fields so the simplest expression is a product of variables A_μ and j^μ. Second, Lagrange's function density, which yields H_{int}, is relativistically invariant and has form $-A_\mu j^\mu$. When introducing this interaction, our theory remains Lorentz invariant. Third, the presence of additional terms in the Hamiltonian that contain the derivatives of our variables is known to have no practical value such that with sufficient certainty, we can restrict ourselves to the density of the interaction in the form $A_\mu j^\mu$.

Summing Hamilton's functions, we obtain the Hamiltonian of the system of interacting fields,

$$H = H_e + H_f + H_{\text{int}}$$

or, in an explicit form,

$$H = \int \psi^+ \left[\alpha^i\left(-i\hbar\frac{\partial}{\partial x^i} - eA_i\right) + (\beta - eA_0)\right]\psi \mathrm{d}\mathbf{r} - \frac{1}{8\pi}\int\left(v_\mu v^\mu - A_{\mu,i}A^{\mu,i}\right)\mathrm{d}\mathbf{r},$$

in which we use the expression for the four-vector of current density,

$$j^\mu = (\rho_e, \mathbf{j}_e) = (-e\psi^+\psi, -e\psi^+\boldsymbol{\alpha}\psi);$$

note that

$$j_\mu = (-e\psi^+\psi, e\psi^+\boldsymbol{\alpha}\psi) = (-e\psi^+\psi, -e\psi^+\alpha_i\psi).$$

Let us verify our Hamiltonian, having written the corresponding equations of motion for variables ψ and A_μ. So, as

$$[\psi(\mathbf{r}), \psi(\mathbf{r}')]_+ = 0$$

and

$$[\psi(\mathbf{r}), \psi^+(\mathbf{r}')]_+ = \delta(\mathbf{r} - \mathbf{r}'),$$

then

$$i\hbar \frac{\partial}{\partial x_0}\psi = [\psi, H] = [\psi, H_e + H_{int}]$$

$$= \int [\psi(\mathbf{r}), \psi^+(\mathbf{r}')]_+ \left(\alpha^i\left(-i\hbar\frac{\partial}{\partial x^i} - eA_i\right) + (\beta - eA_0)\right)\psi(\mathbf{r}')d\mathbf{r}'$$

$$= \left(\alpha^i\left(-i\hbar\frac{\partial}{\partial x^i} - eA_i\right) + (\beta - eA_0)\right)\psi(\mathbf{r}),$$

which is the exact Dirac equation in an external field. Further, applying Hamilton's equations for a scalar field, we have

$$i\hbar\frac{\partial}{\partial x_0}A_\mu = [A_\mu, H] = [A_\mu, H_f] = i\hbar v_\mu,$$

and also

$$i\hbar\frac{\partial}{\partial x_0}v_\mu = [v_\mu, H_f] + [v_\mu, H_{int}] = -i\hbar A_{\mu,i}{}^{,i} + i\hbar \cdot 4\pi j_\mu;$$

i.e.,

$$A_{\mu,\nu}{}^{,\nu} = 4\pi j_\mu,$$

which is the wave equation of Maxwell's classical theory of electromagnetic fields. Thus, our Hamiltonian has classical roots and is quite suitable to describe the interaction in quantum electrodynamics.

Let us separately discuss the Dirac equation in the presence of an electromagnetic field. In this case,

$$i\hbar\dot\psi = (-\alpha^i(p_i + eA_i) + (\beta - eA_0))\psi,$$

in which $p_i = i\hbar\partial/\partial x^i$. As we see, the introduction of an interaction between an electron and an electromagnetic field is accompanied by the traditional replacement,

$$p_\mu \to p_\mu + eA_\mu.$$

It is thereby convenient to separate the external field into static and temporally dependent parts. The former is described, by definition, by potentials $U_\mu(\mathbf{r})$, whereas for the latter we preserve, as before, variables A_μ, such that

$$p_\mu \to p_\mu + eU_\mu + eA_\mu.$$

As we intend to consider a nonstatic interaction according to the perturbation theory, one should rewrite, with another meaning, our Hamiltonian. So,

$$H = H_e + H_f + H_{\text{int}},$$

in which H_f is given by the expression as before; operator H_{int}, which is equal to

$$\int A_\mu j^\mu d\mathbf{r},$$

describes the nonstatic interaction and is considered according to the perturbation theory; Hamiltonian H_e additionally includes the potentials of a static electromagnetic field, such that, by definition,

$$H_e = \int \psi^+ H_D \psi d\mathbf{r} = \int \psi^+(-\alpha^i(p_i + eU_i) + (\beta - eU_0))\psi d\mathbf{r}.$$

Taking into account the static field, we introduce the field variables for an electron. For this purpose, we expand each component $\psi_\mathcal{A}$ with respect to some complete system of operators φ_n,

$$\psi_\mathcal{A}(\mathbf{r}) = \sum_n \langle \mathbf{r}\mathcal{A}|n\rangle \varphi_n,$$

in which expansion coefficients $\langle \mathbf{r}\mathcal{A}|n\rangle$ are some functions of $\mathbf{r}$; via $\mathcal{A}$, we distinguish the four components of a spinor. As $\langle \mathbf{r}\mathcal{A}|n\rangle$, we have the eigenfunctions of Dirac's Hamiltonian H_D:

$$\sum_\mathcal{B} (H_D)_{\mathcal{AB}} \langle \mathbf{r}\mathcal{B}|n\rangle = E_n \langle \mathbf{r}\mathcal{A}|n\rangle,$$

in which E_n are the eigenvalues of operator H_D. Of course, functions $\langle \mathbf{r}\mathcal{A}|n\rangle$ are made orthonormal, such that

$$\langle n|n'\rangle = \sum_\mathcal{A} \int \langle n|\mathbf{r}\mathcal{A}\rangle \langle \mathbf{r}\mathcal{A}|n'\rangle d\mathbf{r} = \delta_{nn'},$$

and satisfy the condition,

$$\sum_n \langle \mathbf{r}\mathcal{A}|n\rangle \langle n|\mathbf{r}'\mathcal{B}\rangle = \delta_{\mathcal{A}\mathcal{B}}\delta(\mathbf{r}-\mathbf{r}').$$

We express φ_n as

$$\varphi_n = \sum_{\mathcal{A}} \int \langle n|\mathbf{r}\mathcal{A}\rangle \psi_{\mathcal{A}}(\mathbf{r}) d\mathbf{r};$$

we obtain analogously for a Hermitian conjugate operator,

$$\varphi_n^+ = \sum_{\mathcal{A}} \int \psi_{\mathcal{A}}^+(\mathbf{r})\langle \mathbf{r}\mathcal{A}|n\rangle d\mathbf{r}.$$

Consequently, through the anticommutation relations

$$\left[\psi_{\mathcal{A}}(\mathbf{r}), \psi_{\mathcal{B}}^+(\mathbf{r}')\right]_+ = \delta_{\mathcal{A}\mathcal{B}} \cdot \delta(\mathbf{r}-\mathbf{r}')$$

and

$$\left[\psi_{\mathcal{A}}(\mathbf{r}), \psi_{\mathcal{B}}(\mathbf{r}')\right]_+ = \left[\psi_{\mathcal{A}}^+(\mathbf{r}), \psi_{\mathcal{B}}^+(\mathbf{r}')\right]_+ = 0,$$

we have

$$\left[\varphi_n, \varphi_{n'}^+\right]_+ = \delta_{nn'}$$

and

$$\left[\varphi_n, \varphi_{n'}\right]_+ = \left[\varphi_n^+, \varphi_{n'}^+\right]_+ = 0.$$

In terms of the new variables, Hamiltonian H_e acquires a simple form,

$$H_e = \sum_{\mathcal{A}\mathcal{B}} \int \psi_{\mathcal{A}}^+(\mathbf{r})(H_D)_{\mathcal{A}\mathcal{B}}\psi_{\mathcal{B}}(\mathbf{r})d\mathbf{r} = \sum_{\mathcal{A}\mathcal{B}} \int \sum_{nn'} \varphi_n^+ \langle n|\mathbf{r}\mathcal{A}\rangle (H_D)_{\mathcal{A}\mathcal{B}}\langle \mathbf{r}\mathcal{B}|n'\rangle \varphi_{n'} d\mathbf{r}$$
$$= \sum_n E_n \varphi_n^+ \varphi_n.$$

We see that for $E_n > 0$, quantity φ_n^+ is a creation operator whereas φ_n is a destruction operator of an electron in state $|n\rangle$. If $E_n < 0$, then φ_n^+ creates whereas φ_n destroys an electron with negative energy; it is equivalent to, respectively, destruction and creation of a positron in state $|n\rangle$.

We proceed to represent our Hamiltonian in a form convenient for calculations according to perturbation theory. We have

$$H = H_0 + H_{\text{int}};$$

here,

$$H_0 = H_e + H_f,$$

in which H_e and H_f are expressible, in a trivial manner, through the corresponding creation and destruction operators of electron–positron and electromagnetic fields. To work with this Hamiltonian, it is convenient to proceed to a so-called interaction picture; this representation is sometimes called the Dirac picture.

Let us consider some physical quantity L, which is a constant of motion, as a function of time t and dynamical variables that are taken at the same moment in time t. Through Heisenberg's equation of motion for a constant of motion, $dL/dt = 0$, hence,

$$i\hbar \frac{dL}{dt} = i\hbar \frac{\partial L}{\partial t} + [L, H] = 0,$$

therefore

$$i\hbar \frac{\partial L}{\partial t} = [H, L].$$

Further, with the aid of a unitary transformation we proceed to the interaction picture, with

$$L' = e^{iH_0 t/\hbar} L e^{-iH_0 t/\hbar}.$$

Differentiating L' with respect to time, we find

$$\frac{\partial L'}{\partial t} = e^{iH_0 t/\hbar} \left(\frac{i}{\hbar} H_0 L + \frac{\partial L}{\partial t} - \frac{i}{\hbar} L H_0 \right) e^{-iH_0 t/\hbar};$$

however, $\partial L/\partial t = [H, L]/i\hbar$ and $H = H_0 + H_{\text{int}}$; consequently,

$$\frac{\partial L'}{\partial t} = \frac{1}{i\hbar} e^{iH_0 t/\hbar} ((H - H_0)L - L(H - H_0)) e^{-iH_0 t/\hbar},$$

hence

$$i\hbar \frac{\partial L'}{\partial t} = [H'_{\text{int}}, L'],$$

in which $H'_{\text{int}} = e^{iH_0t/\hbar}H_{\text{int}}e^{-iH_0t/\hbar}$. The obtained equation of motion defines the temporal variation of a physical quantity in the interaction picture. We see that instead of the total Hamiltonian, H'_{int} arises here; hence, the self-energy of electronic and electromagnetic fields has no influence on any variation in time.

The solution for L' according to the perturbation theory is of special interest. If we consider H_{int} as a perturbation, then one might search for L' in a form of an expansion in a power series in terms of a parameter that characterizes a smallness of Hamiltonian H_{int}. We assume, by definition,

$$L' = L'_0 + L'_1 + L'_2 + L'_3 + \cdots$$

and substitute this expression into the equation of motion and equate the quantities of the same order of smallness on the left and on the right. As a result,

$$\frac{\partial L'_0}{\partial t} = 0, \quad i\hbar\frac{\partial L'_1}{\partial t} = \left[H'_{\text{int}}, L'_0\right], \quad i\hbar\frac{\partial L'_2}{\partial t} = \left[H'_{\text{int}}, L'_1\right], \ldots$$

This is the general solution according to the perturbation theory. Note that L'_0 is explicitly independent of time; quantity L'_0 represents a function of only dynamical variables.

Solution according to perturbation theory

We seek a solution for a *one-electron* problem of quantum electrodynamics, restricting ourselves to the perturbation theory in its second order. In discussing the one-electron problem, we do not err. It turns out that a consideration of a problem of an interaction, for instance, of two electrons, is strongly distinguished from the scheme according to which we proceed. This disparity arises mainly through the appearance of a Coulombic interaction, which we cannot consider as a perturbation. If the fields are characterized by forces that decay more rapidly than an inverse square of a distance, then the outlined scenario entirely preserves its workability. Such an approach is applicable primarily to the one-electron problem, in which the energy of interaction H_{int} plays the role of a perturbation.

We arrange to solve our problem in the interaction picture. Hence, one must choose physical quantity L, which we intend to consider, and proceed to the interaction picture. In the absence of H_{int}, we have one electron, for instance, in state $|\ell\rangle$; therefore, it is natural to choose, as L'_0, the operator for the creation of an electron with energy $E_\ell > 0$, and, taking the interaction into account, to consider its variation with time. Because φ_ℓ^+ is explicitly independent of time, the first equation $\partial L'_0/\partial t = 0$ of perturbation theory is satisfied.

The choice of L is somewhat arbitrary: we might take any physical quantity of interest that is a constant of motion. Our natural interest is regarding the creation operator is applicable to an electron in some state because, in Heisenberg's picture, it characterizes the alteration of the state itself. In contrast, perturbation operator H_{int} is known to us. Its expression remains to be defined in the interaction picture, before the integration of Heisenberg's equations.

By definition,

$$H_{\text{int}} = \int A_\mu j^\mu d\mathbf{r} = -e \int \psi^+ A_\mu \alpha^\mu \psi d\mathbf{r},$$

in which $\alpha^\mu = (\alpha^0, \boldsymbol{\alpha})$; α^0 is a 4×4 unit matrix. Expressing A_μ and ψ through the field variables, we have

$$A_\mu(\mathbf{r}) = \int \left(A_{\mu k} e^{-i\mathbf{k}\cdot\mathbf{r}} + \overline{A}_{\mu k} e^{i\mathbf{k}\cdot\mathbf{r}} \right) d\mathbf{k} \equiv 2 \int A_{\mu k}^{(a)} e^{-i\mathbf{k}\cdot\mathbf{r}} d\mathbf{r}$$

and

$$\psi_A(\mathbf{r}) = \sum_n \langle \mathbf{r} A | n \rangle \varphi_n,$$

in which we apply Dirac's abbreviation,

$$A_{\mu k}^{(a)} = \frac{1}{2}\left(A_{\mu k}^{(1)} + A_{\mu k}^{(-1)} \right) \equiv \frac{1}{2}(A_{\mu k} + \overline{A}_{\mu(-k)}); \quad a = \pm 1.$$

We rewrite H_{int} in the form

$$H_{\text{int}} = -2e \sum_{AB} \sum_{nn'} \int\int \varphi_n^+ \langle n|\mathbf{r}A\rangle (\alpha^\mu e^{-i\mathbf{k}\cdot\mathbf{r}})_{AB} \langle \mathbf{r}B|n'\rangle \varphi_{n'} A_{\mu k}^{(a)} d\mathbf{k}\, d\mathbf{r}$$

$$= -2e \sum_{nn'} \int \varphi_n^+ \varphi_{n'} A_{\mu k}^{(a)} \langle n|\alpha^\mu e^{-i\mathbf{k}\cdot\mathbf{r}}|n'\rangle d\mathbf{k}.$$

We must find quantity $e^{iH_0 t/\hbar} H_{\text{int}} e^{-iH_0 t/\hbar}$; for this purpose, we note that an entire procedure is reduced to this calculation

$$e^{iH_0 t/\hbar} \varphi_n^+ \varphi_{n'} A_{\mu k}^{(a)} e^{-iH_0 t/\hbar}$$

$$= \left(e^{iH_0 t/\hbar} \varphi_n^+ e^{-iH_0 t/\hbar} \right) \cdot \left(e^{iH_0 t/\hbar} \varphi_{n'} e^{-iH_0 t/\hbar} \right) \cdot \left(e^{iH_0 t/\hbar} A_{\mu k}^{(a)} e^{-iH_0 t/\hbar} \right).$$

Hence, it is sufficient to determine the expressions for the field variables in the interaction picture.

Beginning with $A_{\mu k}$, we calculate commutator $[A_{\mu k}, H_0]$; evidently,

$$[A_{\mu k}, H_0] = [A_{\mu k}, H_f] = \left[A_{\mu k}, (-4\pi^2)\int |\mathbf{k}'|^2 A_{\nu k'} \overline{A}^{\nu}_{k'} d\mathbf{k}'\right]$$

$$= -4\pi^2 \int |\mathbf{k}'|^2 A_{\nu k'} [A_{\mu k}, \overline{A}^{\nu}_{k'}] d\mathbf{k}'$$

$$= -4\pi^2 \int |\mathbf{k}'|^2 A_{\nu k'} \left(\frac{\hbar g^{\nu}_{\mu}}{4\pi^2 |\mathbf{k}'|} \delta(\mathbf{k} - \mathbf{k}')\right) d\mathbf{k}' = -\hbar |\mathbf{k}| A_{\mu k},$$

i.e.,

$$A_{\mu k} H_0 = (H_0 - \hbar |\mathbf{k}|) A_{\mu k}.$$

Using this operator relation, we proceed to the interaction picture:

$$A'_{\mu k} = e^{iH_0 t/\hbar} A_{\mu k} e^{-iH_0 t/\hbar} = e^{iH_0 t/\hbar} A_{\mu k} \left(\sum_j \frac{1}{j!} \left(-\frac{it}{\hbar}\right)^j H_0^j\right)$$

$$= e^{iH_0 t/\hbar} \left(\sum_j \frac{1}{j!} \left(-\frac{it}{\hbar}\right)^j (H_0 - \hbar|\mathbf{k}|)^j\right) A_{\mu k} = e^{iH_0 t/\hbar} e^{-i(H_0 - \hbar|\mathbf{k}|)t/\hbar} A_{\mu k} = e^{i|\mathbf{k}|t} A_{\mu k}.$$

Analogously, we find $\overline{A}'_{\mu k} = e^{-i|\mathbf{k}|t} \overline{A}_{\mu k}$. Consequently,

$$A^{(a)'}_{\mu k} = \frac{1}{2}\left(A'_{\mu k} + \overline{A}'_{\mu(-k)}\right) = \frac{1}{2}\left(A^{(1)}_{\mu k} e^{i|\mathbf{k}|t} + A^{(-1)}_{\mu k} e^{-i|\mathbf{k}|t}\right) = A^{(a)}_{\mu k} e^{ia|\mathbf{k}|t};$$

here, as before, we imply the average summation with respect to a. Note that if, in some expression, quantity a appears linearly, $a = (1 + (-1))/2 = 0$. Hence, $a^2 = 1$.

Applying the same scenario, we determine φ'_n. We have

$$[\varphi_n, H_0] = [\varphi_n, H_e] = \sum_{n'} E_{n'} [\varphi_n, \varphi^+_{n'} \varphi_{n'}] = E_n \varphi_n,$$

therefore

$$\varphi_n H_0 = (H_0 + E_n)\varphi_n;$$

consequently,

$$\varphi'_n = e^{iH_0 t/\hbar} \varphi_n e^{-iH_0 t/\hbar} = e^{-iE_n t/\hbar} \varphi_n.$$

Analogously, $\varphi^{+'}_n = e^{iE_n t/\hbar} \varphi^+_n$.

Thus, one might conclude that

$$H'_{\text{int}} = -2e \sum_{nn'} \int \varphi_n^+ \varphi_{n'} A_{\mu k}^{(a)} \langle n|\alpha^\mu e^{-i\mathbf{k}\cdot\mathbf{r}}|n'\rangle e^{i(E_n - E_{n'} + a\hbar|\mathbf{k}|)t/\hbar} \, \mathbf{dk}$$

is the sought expression for H_{int} in the interaction picture. We proceed to apply the general formulae of perturbation theory of second order, which we obtained in the preceding section; so, $L_0' = \varphi_\ell^+$ and $\partial \varphi_\ell^+/\partial t = 0$. From equation $i\hbar \partial L_1'/\partial t = [H'_{\text{int}}, L_0']$, we calculate $L_1' = \varphi_{\ell(1)}^+$,

$$i\hbar \frac{\partial}{\partial t} \varphi_{\ell(1)}^+ = -2e \sum_{nn'} \int [\varphi_n^+ \varphi_{n'}, \varphi_\ell^+] A_{\mu k}^{(a)} \langle n|\alpha^\mu e^{-i\mathbf{k}\cdot\mathbf{r}}|n'\rangle e^{i(E_n - E_{n'} + a\hbar k)t/\hbar} \, \mathbf{dk},$$

in which, for the sake of convenience, we use $|\mathbf{k}| = k$. Taking into account that

$$[\varphi_n^+ \varphi_{n'}, \varphi_\ell^+] = \varphi_n^+ \varphi_{n'} \varphi_\ell^+ - \varphi_\ell^+ \varphi_n^+ \varphi_{n'} = \varphi_n^+ [\varphi_{n'}, \varphi_\ell^+]_+ = \varphi_n^+ \delta_{n'\ell},$$

we have

$$i\hbar \frac{\partial}{\partial t} \varphi_{\ell(1)}^+ = -2e \sum_n \int \varphi_n^+ A_{\mu k}^{(a)} \langle n|\alpha^\mu e^{-i\mathbf{k}\cdot\mathbf{r}}|\ell\rangle e^{i(E_n - E_\ell + a\hbar k)t/\hbar} \, \mathbf{dk}.$$

Integrating, we eventually find

$$\varphi_{\ell(1)}^+ = 2e \sum_n \int \varphi_n^+ A_{\mu k}^{(a)} \frac{\langle n|\alpha^\mu e^{-i\mathbf{k}\cdot\mathbf{r}}|\ell\rangle}{E_n - E_\ell + a\hbar k} e^{i(E_n - E_\ell + a\hbar k)t/\hbar} \, \mathbf{dk}.$$

This first correction to the creation operator of the electron is caused by the interaction with an electromagnetic field.

Regarding second correction $\varphi_{\ell(2)}^+$, the situation is slightly complicated; this correction is defined through equation $i\hbar \partial \varphi_{\ell(2)}^+/\partial t = \left[H'_{\text{int}}, \varphi_{\ell(1)}^+\right]$, i.e.,

$$i\hbar \frac{\partial}{\partial t} \varphi_{\ell(2)}^+ = -4e^2 \sum_{nn'n''} \iint \left[\varphi_{n'}^+ \varphi_{n''} A_{\nu k'}^{(b)}, \varphi_n^+ A_{\mu k}^{(a)}\right] \langle n'|\alpha^\nu e^{-i\mathbf{k}'\cdot\mathbf{r}}|n''\rangle \cdot$$
$$\cdot \frac{\langle n|\alpha^\mu e^{-i\mathbf{k}\cdot\mathbf{r}}|\ell\rangle}{E_n - E_\ell + a\hbar k} e^{i(E_{n'} - E_{n''} + E_n - E_\ell + b\hbar k' + a\hbar k)t/\hbar} \, \mathbf{dk} \, \mathbf{dk'}.$$

A legitimate question arises: in what order should we locate the field operators? A result depends on this order.

Normal product

Considering the second quantization of a fermion field, we have noted that, according to convention, all creation operators must be located to the left with respect to destruction operators. This is the so-called *normal* product of the field variables. This procedure is essential because we can represent any state vector in a form of some product of creation operators that act on the vector of vacuum state $|0\rangle$ that is determined at a fixed moment in time. Because only these states represent a physical interest, we operate with only such states. If, in a process of seeking a solution, some destruction operator arises on the left, then one must simply apply the commutation or anticommutation relations and perform normal ordering. As a result, we again obtain some expansion in a power series in terms of the field operators, but a part of them consists entirely of the creation operators. Other combinations that contain the destruction operators on the right produce exactly zero when acting on the vacuum vector. This procedure is logical. In quantum mechanics, we work with operators for which the order of location is generally important; therefore, it is natural to define at least some convention regarding the order of location of operators to enable a comparison of theory and experiment. In the literature, one might sometimes meet a more prosaic argument. A normal form is a convenient form to calculate the matrix elements.

We proceed to transform the obtained commutator,

$$\left[\varphi_{n'}^+\varphi_{n''}A_{\nu k'}^{(b)}, \varphi_n^+ A_{\mu k}^{(a)}\right] = \varphi_{n'}^+\varphi_{n''}\varphi_n^+ A_{\nu k'}^{(b)}A_{\mu k}^{(a)} - \varphi_n^+\varphi_{n'}^+\varphi_{n''}A_{\mu k}^{(a)}A_{\nu k'}^{(b)}$$

in the expression for $i\hbar\partial\varphi_{\ell(2)}^+/\partial t$; one should perform normal ordering of our variables.

Initially, we consider φ-variables that satisfy the anticommutation relations by unity; φ_n^+ is a creation operator for $E_n > 0$ but for $E_n < 0$, φ_n is the creation operator. One might readily understand that

$$\varphi_n\varphi_{n'}^+ = (Np) + \delta_{nn'}\eta(E_n),$$

in which symbol (Np) signifies the normal product and $\eta(E_n)$ is the Heaviside function. Recall that $\eta(E_n > 0) = 1$ and $\eta(E_n < 0) = 0$. If our variables become interchanged, then

$$\varphi_{n'}^+\varphi_n = (Np) + \delta_{nn'}\eta(-E_n).$$

Let us consider variables $A_{\mu k}$ that satisfy commutation relation

$$[A_{\mu k}, \overline{A}_{\nu k'}] = \hbar g_{\mu\nu}\delta(\mathbf{k} - \mathbf{k'})/4\pi^2 k,$$

in which $A_{\mu k}$ is a creation operator and $\overline{A}_{\nu k'}$ is a destruction operator. Assuming

$$A_{\mu k} \equiv A_{\mu k}^{(1)} \text{ and } \overline{A}_{\nu k'} \equiv A_{\nu(-k')}^{(-1)},$$

we obtain

$$A_{\mu k}^{(-1)}A_{\nu k'}^{(1)} = (Np) - \frac{\hbar g_{\mu\nu}}{4\pi^2 k}\delta(\mathbf{k} + \mathbf{k'}).$$

Applying Dirac's abbreviation, we convert this expression to an elegant form,

$$A^{(a)}_{\mu k} A^{(b)}_{\nu k'} = (Np) - \frac{\hbar g_{\mu\nu}}{4\pi^2 k} \delta(\mathbf{k} + \mathbf{k}') \frac{(1-a)(1+b)}{4};$$

here, a and $b = \pm 1$.

We have sufficient knowledge to transform the commutator. So,

$$\varphi^+_{n'} \varphi_{n''} \varphi^+_n = ((Np) + \delta_{n'n''} \eta(-E_{n'})) \varphi^+_n = (Np) + \varphi^+_{n'} \delta_{nn''} \eta(E_n) + \varphi^+_n \delta_{n'n''} \eta(-E_{n'}),$$

in which we take into account that

$$(Np)\varphi^+_n \to \varphi^+_{n'}(\varphi_{n''}\varphi^+_n) = \varphi^+_{n'}((Np) + \delta_{nn''}\eta(E_n)).$$

Analogously,

$$\varphi^+_n \varphi^+_{n'} \varphi_{n''} = \varphi^+_n((Np) + \delta_{n'n''}\eta(-E_{n''})) = (Np) - \varphi^+_{n'}\delta_{nn''}\eta(-E_{n''}) + \varphi^+_n \delta_{n'n''}\eta(-E_{n''}).$$

The situation is simpler for quantities $A^{(a)}_{\mu k}$, because

$$A^{(b)}_{\nu k'} A^{(a)}_{\mu k} = (Np) - \frac{\hbar g_{\nu\mu}}{4\pi^2 k} \delta(\mathbf{k} + \mathbf{k}') \frac{(1-b)(1+a)}{4} \equiv (Np) - \frac{\hbar g_{\mu\nu}}{4\pi^2 k} \delta(\mathbf{k} + \mathbf{k}') \frac{(1+a)(a-b)}{4};$$

also,

$$A^{(a)}_{\mu k} A^{(b)}_{\nu k'} = (Np) - \frac{\hbar g_{\mu\nu}}{4\pi^2 k} \delta(\mathbf{k} + \mathbf{k}') \frac{(1-a)(b-a)}{4},$$

in which we apply an algebraic identity when calculating

$$(1-a)(1+b) = (1-a)(b-a) = (b-a)(1+b),$$

which is valid for the case a and $b = \pm 1$. This validity is directly verifiable.

Thus,

$$\left[\varphi^+_{n'}\varphi_{n''}A^{(b)}_{\nu k'}, \varphi^+_n A^{(a)}_{\mu k}\right] = (Np) - \frac{\hbar g_{\mu\nu}}{4\pi^2 k} \delta(\mathbf{k} + \mathbf{k}') \frac{(a-b)}{4} \cdot$$
$$\cdot \left[(1+a)\left(\varphi^+_{n'}\delta_{nn''}\eta(E_n) + \varphi^+_n \delta_{n'n''}\eta(-E_{n'})\right)\right.$$
$$\left. + (1-a)\left(\varphi^+_n \delta_{n'n''}\eta(-E_{n''}) - \varphi^+_{n'}\delta_{nn''}\eta(-E_{n''})\right)\right].$$

In this expansion, we are primarily interested in the field operators to smaller powers, such that

$$\left[\varphi_{n'}^+\varphi_{n''}A_{\nu k'}^{(b)}, \varphi_n^+ A_{\mu k}^{(a)}\right] = -\frac{\hbar g_{\mu\nu}}{16\pi^2 k}\delta(\mathbf{k}+\mathbf{k}')(a-b)\cdot$$
$$\cdot \left[\varphi_{n'}^+ \delta_{nn''}((1+a)\eta(E_n) - (1-a)\eta(-E_n)) + 2\varphi_n^+ \delta_{n'n''}\eta(-E_{n'})\right] + \text{higher powers of } \varphi.$$

Restricting ourselves to the first explicitly written approximation, we return to $i\hbar\partial\varphi_{\ell(2)}^+/\partial t$. We first perform the integration with respect to $\mathbf{k}'$; this action simply leads to a substitution of $\mathbf{k}'$ by $-\mathbf{k}$. We note that the argument of the exponential function contains sum $a+b$, which is multiplied by ikt; this sum is equal to zero. We perform the summation with respect to n''; as a result,

$$i\hbar\frac{\partial}{\partial t}\varphi_{\ell(2)}^+ = \frac{\hbar e^2}{4\pi^2}(a-b)\sum_{nn'}\int \varphi_{n'}^+ \frac{\langle n'|\alpha_\mu e^{i\mathbf{k}\cdot\mathbf{r}}|n\rangle}{k}\frac{\langle n|\alpha^\mu e^{-i\mathbf{k}\cdot\mathbf{r}}|\ell\rangle}{E_n - E_\ell + a\hbar k}\cdot$$
$$\cdot ((1+a)\eta(E_n) - (1-a)\eta(-E_n))e^{i(E_{n'}-E_\ell)t/\hbar}\mathbf{dk} +$$
$$+ \frac{\hbar e^2}{4\pi^2}(a-b)\sum_{nn'}\int \varphi_n^+ \frac{\langle n'|\alpha_\mu e^{i\mathbf{k}\cdot\mathbf{r}}|n'\rangle}{k}\frac{\langle n|\alpha^\mu e^{-i\mathbf{k}\cdot\mathbf{r}}|\ell\rangle}{E_n - E_\ell + a\hbar k}2\eta(-E_{n'})e^{i(E_n - E_\ell)t/\hbar}\mathbf{dk}.$$

We perform the summation with respect to b; this action yields the substitution of difference $a - b$ by a; eventually,[13]

$$i\hbar\frac{\partial}{\partial t}\varphi_{\ell(2)}^+ = \frac{e^2}{\pi}\sum_n \varphi_n^+ (Y_{n\ell} + Z_{n\ell})e^{i(E_n - E_\ell)t/\hbar},$$

in which

$$Y_{n\ell} = \frac{\hbar}{4\pi}\sum_{n'}\int a\frac{\langle n|\alpha_\mu e^{i\mathbf{k}\cdot\mathbf{r}}|n'\rangle}{k}\frac{\langle n'|\alpha^\mu e^{-i\mathbf{k}\cdot\mathbf{r}}|\ell\rangle}{E_{n'} - E_\ell + a\hbar k}((1+a)\eta(E_{n'}) - (1-a)\eta(-E_{n'}))\mathbf{dk}$$

and

$$Z_{n\ell} = \frac{\hbar}{2\pi}\sum_{n'}\int a\frac{\langle n'|\alpha_\mu e^{i\mathbf{k}\cdot\mathbf{r}}|n'\rangle}{k}\frac{\langle n|\alpha^\mu e^{-i\mathbf{k}\cdot\mathbf{r}}|\ell\rangle}{E_n - E_\ell + a\hbar k}\eta(-E_{n'})\mathbf{dk}.$$

These expressions are our principal working formulae. Let us briefly discuss their physical meaning. What do matrices Y and Z represent?

We produce a heuristic analysis to correlate the developed formalism of the field theory with the traditional description in the Schrödinger picture. For this purpose, we replace operator ψ with vector $|\psi\rangle$, and expansion

$$\psi_A = \sum_n \langle \mathbf{r} A | n \rangle \varphi_n$$

with expansion

$$|\psi\rangle = \sum_n \varphi_n(t) e^{-iE_n t/\hbar} |n\rangle$$

in terms of eigenfunctions $e^{-iE_n t/\hbar} \langle \mathbf{r} | n \rangle$ of stationary states; φ_n become the ordinary coefficients. We substitute this expansion into Schrödinger's equation $i\hbar \partial |\psi\rangle / \partial t = (H_0 + H_{\text{int}}) |\psi\rangle$; as a result, we obtain

$$i\hbar \frac{\partial \varphi_{n'}}{\partial t} = \sum_n \varphi_n \langle n' | H_{\text{int}} | n \rangle e^{i(E_{n'} - E_n)t/\hbar},$$

in which

$$H_0 |n\rangle = E_n |n\rangle.$$

One might further introduce the corrections for calculations according to perturbation theory, with

$$\varphi_n = \varphi_{n(0)} + \varphi_{n(1)} + \varphi_{n(2)} + \cdots;$$

the diagonal elements in the expansions of

$$i\hbar \partial \varphi_{n(1)} / \partial t, \; i\hbar \partial \varphi_{n(2)} / \partial t, \ldots$$

in terms of φ_n correspond to a variation of energy. For instance, in the first order, this variation is simply equal to diagonal matrix element $\langle n | H_{\text{int}} | n \rangle$ of the perturbation. In quantum theory, $i\hbar \partial / \partial t$ is the operator of energy for which the diagonal matrix elements represent its variation.

Matrix elements $Y_{n\ell}$ and $Z_{n\ell}$, which are accurate within a factor e^2/π, are analogous to the matrix elements of a perturbation in the Schrödinger picture. Quantity

$$\frac{e^2}{\pi}(Y_{\ell\ell} + Z_{\ell\ell})$$

is therefore interpreted as the variation of energy of our electron in an electromagnetic field. Note that this is only an increment of energy, not an absolute value. Taking into account the interaction, matrices Y and Z describe the variation of the creation operator. Only they are necessary to calculate the radiative corrections of the second order.

Ultraviolet divergences

We have obtained the general expressions for quantities Y and Z that define the energy variation, which is caused by an interaction between an electron and an electromagnetic field. These quantities eventually become reduced to integrals with respect to k_1, k_2, and k_3. Scrutiny indicates that these integrals are divergent because of integrands that decrease slowly in the region of large values of k. Such divergences, which exist according to both the Schrödinger and Heisenberg pictures, bear a general name — ultraviolet divergences. The essence of the problem is hidden in the choice of the Hamiltonian; if the Hamiltonian remains unaltered, it is difficult to hope for deliverance from all divergent expressions in principle. Nevertheless, there exists a way forward — the subsidiary *regularizations* of the resulting integrals, but not the naive possibility of discarding the infinite expressions from consideration.

The simplest procedure of regularization is reduced to a truncation of the integrals at some concrete value $k = \chi$; recall that $k \equiv |\mathbf{k}|$. On choosing a sufficiently large finite value for χ, one might obtain a quite convergent result. However, the cost of such a regularization is a loss of relativistic invariance because, when integrating, the procedure of truncation of the region of large values of k fails to be Lorentz invariant. We thus automatically restrict the possibilities of our Hamiltonian and we assign quantum electrodynamics to workability only in the region of small energies. However, we must sacrifice something, such as an exclusion of high-energy processes from consideration. Note that this limit is not too small; it amounts to 10^8 eV. One might suppose that the developed quantum theory becomes correct for energies, which are knowingly smaller than the indicated limit, and loses physical meaning in the region of high energies that transcend this limit.

The peculiar unsuitability of such quantum electrodynamics in the region of high energies naturally emphasizes a true limitation of the entire theory. The theory itself cannot elucidate all processes, and we are consequently forced to perform regularization. A poor excuse might only reflect the fact that, in a framework of calculations according to the perturbation theory, all approaches to quantum field theory experience ultraviolet divergences. The procedure of regularization of interaction energy must be a natural procedure.

Regularization of interaction energy

In addition to a trivial truncation, there are other methods to regularize our integrals. Let us consider, for instance, such a method. We restrict the expressions in the denominators of Y and Z; namely, we use

$$|E_n| + |E_\ell| + \hbar k < 2\hbar\chi,$$

in which factor 2 is introduced for convenience. We thus impose a condition on a sum of absolute values of energies E_n, E_ℓ, and $\hbar k$, not simply on the values of k, as in the case of a truncation. To better understand this distinction, one must compare, for instance, the calculation of Y with the aid of a method of regularization.

To proceed, we first slightly simplify the general expression for $Y_{n\ell}$, applying the known properties of the Heaviside function. So,

$$\eta(E_n) + \eta(-E_n) = 1 \text{ and } \eta(E_n) - \eta(-E_n) = \frac{E_n}{|E_n|};$$

consequently,

$$(1+a)\eta(E_{n'}) - (1-a)\eta(-E_{n'}) = a + \frac{E_{n'}}{|E_{n'}|}.$$

Further, noting that $a^2 = 1$, we rewrite the expression for $Y_{n\ell}$ in the form

$$Y_{n\ell} = \frac{\hbar}{4\pi} \sum_{n'} \int \frac{\langle n|\alpha_\mu e^{i\mathbf{k}\cdot\mathbf{r}}|n'\rangle \langle n'|\alpha^\mu e^{-i\mathbf{k}\cdot\mathbf{r}}|\ell\rangle}{aE_{n'} - aE_\ell + \hbar k} \left(a + \frac{E_{n'}}{|E_{n'}|}\right) \frac{d\mathbf{k}}{k}.$$

Bypassing the intermediate summation, we have

$$\langle n|Y|\ell\rangle = \frac{\hbar}{4\pi} \int \langle n|\alpha_\mu e^{i\mathbf{k}\cdot\mathbf{r}} \frac{a + E/|E|}{aE - aE_\ell + \hbar k} \alpha^\mu e^{-i\mathbf{k}\cdot\mathbf{r}}|\ell\rangle \frac{d\mathbf{k}}{k}.$$

We see that the quantity Y represents a definite operator,

$$Y = \frac{\hbar}{4\pi} \int \alpha_\mu e^{i\mathbf{k}\cdot\mathbf{r}} \frac{a + E/|E|}{|E| - aE_\ell + \hbar k} e^{-i\mathbf{k}\cdot\mathbf{r}} \alpha^\mu \frac{d\mathbf{k}}{k},$$

in which, in the denominator, we preliminarily have replaced aE by $|E|$. Such a replacement in the denominator is entirely appropriate, because quantity aE equals either $|E|$ or $-|E|$. If $aE = -|E|$, then $a + E/|E| = 0$, and the expression for Y becomes equal to zero. Finally, for an arbitrary one-electron operator y, by definition,

$$\tilde{y} = e^{-i\mathbf{k}\cdot\mathbf{r}} y e^{i\mathbf{k}\cdot\mathbf{r}};$$

then, we replace the integration variable $\mathbf{k}$ with $-\mathbf{k}$ to obtain

$$Y = \frac{\hbar}{4\pi} \int \alpha_\mu \frac{a + \tilde{E}/|\tilde{E}|}{|\tilde{E}| - aE_\ell + \hbar k} \alpha^\mu \frac{d\mathbf{k}}{k}$$

that is accurate within a sign.

We must operate with this new expression; understanding the meaning of $\tilde{E}$ remains. In general, E is the Dirac operator of energy for an electron taking account of static fields, i.e.,

$$E = \alpha \cdot (\mathbf{p} + e\mathbf{U}) + \beta - eU_0;$$

evidently,

$$\tilde{E} = e^{-i\mathbf{k}\cdot\mathbf{r}} E e^{i\mathbf{k}\cdot\mathbf{r}} = e^{-i\mathbf{k}\cdot\mathbf{r}} \alpha \cdot (\mathbf{p} + e\mathbf{U}) e^{i\mathbf{k}\cdot\mathbf{r}} + \beta - eU_0.$$

Static potential **U** commutes with $e^{i\mathbf{k}\cdot\mathbf{r}}$; for $\mathbf{p}e^{i\mathbf{k}\cdot\mathbf{r}}$, we have

$$e^{i\mathbf{k}\cdot\mathbf{r}}\mathbf{p} - \mathbf{p}e^{i\mathbf{k}\cdot\mathbf{r}} = i\hbar \frac{\partial}{\partial \mathbf{r}}(e^{i\mathbf{k}\cdot\mathbf{r}}) = -\hbar\mathbf{k}e^{i\mathbf{k}\cdot\mathbf{r}}.$$

Therefore,

$$\tilde{E} = \alpha \cdot (\mathbf{P} + \hbar\mathbf{k}) + \beta - eU_0,$$

in which we introduce the designation

$$\mathbf{P} = \mathbf{p} + e\mathbf{U}.$$

We return to the discussion of the procedure of regularization; we involve our conditions $|E| + |E_\ell| + \hbar k < 2\hbar\chi$ and $k < \chi$. This involvement is simply achieved: one must add the corresponding Heaviside function into the integrand of Y, namely,

$$\eta(2\hbar\chi - |E| - |E_\ell| - \hbar k) \text{ or } \eta(\chi - k).$$

As a result,

$$Y_I = \frac{\hbar}{4\pi} \int \alpha_\mu \frac{a + \tilde{E}/|\tilde{E}|}{|\tilde{E}| - aE_\ell + \hbar k} \eta(2\hbar\chi - |\tilde{E}| - |E_\ell| - \hbar k) \alpha^\mu \frac{d\mathbf{k}}{k}$$

and

$$Y_{II} = \frac{\hbar}{4\pi} \int \alpha_\mu \frac{a + \tilde{E}/|\tilde{E}|}{|\tilde{E}| - aE_\ell + \hbar k} \eta(\chi - k) \alpha^\mu \frac{d\mathbf{k}}{k}.$$

Here, we take into account that

$$\tilde{\eta}(2\hbar\chi - |E| - |E_\ell| - \hbar k) = \eta(2\hbar\chi - |\tilde{E}| - |E_\ell| - \hbar k),$$

and, for quantity Y, we separate the expressions according to the method of regularization.

We consider the region of large values of k. Obviously,

$$|\tilde{E}| \equiv (\tilde{E}^2)^{1/2} = (\hbar^2 k^2 + 2(\mathbf{k}\cdot\mathbf{P}) - 2(\boldsymbol{\alpha}\cdot\mathbf{k})eU_0 + \cdots)^{1/2},$$

in which we take into account that $\alpha\beta + \beta\alpha = 0$. Other terms have no interest because k is implied to be great, whereas the unwritten terms do not contain $\mathbf{k}$. Thus, we have

$$|\tilde{E}| = \hbar k + \frac{\mathbf{k}\cdot\mathbf{P}}{k} - \frac{\boldsymbol{\alpha}\cdot\mathbf{k}}{k}eU_0 + o(\chi^{-1}).$$

Furthermore,

$$|\tilde{E}| - aE_\ell + \hbar k \approx 2\hbar k,$$

and

$$\frac{\tilde{E}}{|\tilde{E}|} = \frac{\hbar(\boldsymbol{\alpha}\cdot\mathbf{k}) + \cdots}{\hbar k + \cdots} \approx \frac{\boldsymbol{\alpha}\cdot\mathbf{k}}{k} + o(\chi^{-2}) = \boldsymbol{\alpha}\cdot\mathbf{l},$$

in which $\mathbf{l} = \mathbf{k}/k$. In the chosen approximation, consequently,

$$\frac{a + \tilde{E}/|\tilde{E}|}{|\tilde{E}| - aE_\ell + \hbar k} = \frac{\sum_{a=\pm 1} a/2 + \boldsymbol{\alpha}\cdot\mathbf{l}}{2\hbar k} = \frac{\boldsymbol{\alpha}\cdot\mathbf{l}}{2\hbar k}$$

and

$$\eta(2\hbar\chi - |\tilde{E}| - |E_\ell| - \hbar k) = \eta(2\hbar\chi - 2\hbar k - \mathbf{l}\cdot\mathbf{P} + (\boldsymbol{\alpha}\cdot\mathbf{l})eU_0 - |E_\ell| + \cdots)$$
$$\equiv \eta(\chi - k + F),$$

in which

$$F = \frac{1}{2\hbar}(-\mathbf{l}\cdot\mathbf{P} + (\boldsymbol{\alpha}\cdot\mathbf{l})eU_0 - |E_\ell| + \cdots).$$

We substitute the obtained approximations into the expressions for Y_I and Y_II and calculate their difference,

$$Y_\text{I} - Y_\text{II} = \frac{\hbar}{4\pi}\int \alpha_\mu \frac{\boldsymbol{\alpha}\cdot\mathbf{l}}{2\hbar k}(\eta(\chi - k + F) - \eta(\chi - k))\alpha^\mu \frac{d\mathbf{k}}{k}.$$

It is convenient to proceed to spherical coordinates; in this case,

$$d\mathbf{k} = k^2 dk \cdot \sin\theta\, d\theta\, d\phi$$

and

$$l = (\cos\phi \sin\theta, \; \sin\phi \sin\theta, \; \cos\theta).$$

Hence,

$$Y_I - Y_{II} = \frac{1}{8\pi}\int d\phi \int \alpha_\mu (\alpha \cdot l)(\int (\eta(\chi - k + F) - \eta(\chi - k))dk)\alpha^\mu \sin\theta \, d\theta.$$

The integral of the difference of Heaviside functions with respect to k yields the difference of the arguments of these functions; therefore,

$$Y_I - Y_{II} = \frac{1}{8\pi}\iint \alpha_\mu (\alpha \cdot l)F\alpha^\mu \sin\theta \, d\theta \, d\phi$$

$$= \frac{1}{8\pi}\iint \alpha_\mu \frac{(\alpha \cdot l)}{2\hbar}(-l \cdot \mathbf{P} + (\alpha \cdot l)eU_0$$

$$+ \text{ terms that are independent of } l)\alpha^\mu \sin\theta \, d\theta \, d\phi.$$

If we forget momentarily that we operate with q-numbers, one might perform the integration over the angles. Only the terms that are quadratic with respect to components l^i of vector l yield a non-zero contribution, because

$$\int_0^{2\pi}\int_0^\pi \cos^2\phi \cdot \sin^2\theta \cdot \sin\theta \, d\theta \, d\phi = \frac{4\pi}{3}, \quad \int_0^{2\pi}\int_0^\pi \sin^2\phi \cdot \sin^2\theta \cdot \sin\theta \, d\theta \, d\phi = \frac{4\pi}{3}$$

and

$$\int_0^{2\pi}\int_0^\pi \cos^2\theta \cdot \sin\theta \, d\theta \, d\phi = \frac{4\pi}{3},$$

whereas

$$\int_0^{2\pi}\int_0^\pi l^i l^j \sin\theta \, d\theta \, d\phi = 0 \quad \text{for } i \neq j$$

and

$$\int_0^{2\pi}\int_0^\pi l^i \cdot \sin\theta \, d\theta \, d\phi = 0.$$

As a result,

$$Y_I - Y_{II} = \alpha_\mu \left(-\frac{\alpha \cdot \mathbf{P}}{12\hbar} + \frac{eU_0}{4\hbar}\right)\alpha^\mu.$$

Recalling that $(\alpha^\mu)^2 = 1$, $[\alpha^i, \alpha^0] = 0$ and $[\alpha^i, \alpha^j]_+ = 0$ for $i \neq j$, we have

$$\alpha_\mu \alpha^\mu = (\alpha_0)^2 - \sum_i (\alpha^i)^2 = -2$$

and

$$\alpha_\mu (\boldsymbol{\alpha} \cdot \mathbf{P}) \alpha^\mu = \alpha_0 \alpha^0 (\boldsymbol{\alpha} \cdot \mathbf{P}) + \alpha_1 \alpha^1 (\alpha^1 P^1 - \alpha^2 P^2 - \alpha^3 P^3) + \\ + \alpha_2 \alpha^2 (\alpha^2 P^2 - \alpha^1 P^1 - \alpha^3 P^3) + \alpha_3 \alpha^3 (\alpha^3 P^3 - \alpha^1 P^1 - \alpha^2 P^2) = 2(\boldsymbol{\alpha} \cdot \mathbf{P}).$$

Thus, eventually,

$$Y_I - Y_{II} = -\frac{\boldsymbol{\alpha} \cdot \mathbf{P}}{6\hbar} - \frac{eU_0}{2\hbar}.$$

Note that in these calculations we have assumed that F is an exceptionally diagonal quantity. We have operated in just such a representation because our interest is only the estimated values, not the general nuances regarding the calculations of the integrals of q-numbers. On choosing another representation, the result must remain invariant.

Difference $Y_I - Y_{II}$ is independent of χ and has a finite value, as it should be for the case of sufficiently large values of χ. Which method of regularization is superior? In practice, in various problems of quantum electrodynamics, one might apply various regularizations, which are chosen according to concrete physical reasons. For instance, one might restrict a sum of only kinetic energies, not a sum of total energies; that is, one might assume that

$$|E + eU_0| + |E_\ell + eU_0| + \hbar k < 2\hbar \chi,$$

in which, for the Dirac electron,

$$E = \text{kinetic energy} - eU_0.$$

Together with the procedures that we have considered, this regularization is also quite reasonable.

Radiative corrections

Renormalization of mass

In this chapter, following Dirac's general theory, we proceed to study interacting quantum fields. Beginning with a simple particular case, we consider a variation of energy of an electron in the absence of static fields. In this case, the operator of energy for an electron is sufficiently simple because $U_\mu = 0$. So,

$$E = \boldsymbol{\alpha} \cdot \mathbf{p} + \beta \quad \text{and} \quad |E| = (\mathbf{p}^2 + m^2)^{1/2}.$$

Let us scrutinize how the expressions for Y and Z behave. For such calculations, we must involve some method of regularization. As we see further, even in a trivial case in which a static field is absent, the consequences of regularization require "altering" of an electronic mass. However, the consequences require renormalizing only a theoretical mass in the Dirac equation, not an observable mass. This subtle effect of quantum electrodynamics is nothing more than a "theory in a theory."

We first consider quantity Z; by definition,

$$Z_{n\ell} = \frac{\hbar}{\pi} \int a J_k^\mu \frac{\langle n|\alpha_\mu e^{-i\mathbf{k}\cdot\mathbf{r}}|\ell\rangle}{E_n - E_\ell + a\hbar k} \cdot \frac{d\mathbf{k}}{k}.$$

Here,

$$J_k^\mu = \frac{1}{2}\sum_{n'} \eta(-E_{n'})\langle n'|\alpha^\mu e^{i\mathbf{k}\cdot\mathbf{r}}|n'\rangle = \frac{1}{2}\sum_{n'n''}\langle n'|e^{i\mathbf{k}\cdot\mathbf{r}}|n''\rangle\langle n''|\alpha^\mu \eta(-E)|n'\rangle;$$

when transferring to the right part of this equality, we imply an ordinary matrix product. Moreover,

$$\eta(-E)|n'\rangle = \eta(-E_{n'})|n'\rangle.$$

Let us calculate J_k^μ in a representation in which matrix E is diagonal. The explicit form of E prompts us to the momentum representation. Hence, using

$$\langle \mathbf{r}|n\rangle = \frac{1}{(2\pi\hbar)^{3/2}} e^{i\mathbf{p}\cdot\mathbf{r}/\hbar},$$

we have

$$\langle n'|e^{i\mathbf{k}\cdot\mathbf{r}}|n''\rangle = \frac{1}{(2\pi\hbar)^3}\int e^{i(\mathbf{k}+((\mathbf{p}''-\mathbf{p}')/\hbar))\cdot\mathbf{r}}d\mathbf{r} = \delta(\mathbf{p}'-\mathbf{p}''-\hbar\mathbf{k}).$$

Our sums with respect to n' and n'' convert into integrals with respect to $d\mathbf{p}'$ and $d\mathbf{p}''$. As a result,

$$J_k^\mu = \frac{1}{2(2\pi\hbar)^3}\int \alpha^\mu \eta(-E_{p'})d\mathbf{p}'\int e^{-i\mathbf{k}\cdot\mathbf{r}}d\mathbf{r} \sim \delta(\mathbf{k}),$$

in which we have performed an elementary integration of function $e^{i(\mathbf{p}''-\mathbf{p}')\cdot\mathbf{r}/\hbar}\delta(\mathbf{p}'-\mathbf{p}''-\hbar\mathbf{k})$ with respect to $d\mathbf{p}''$ to obtain function $e^{-i\mathbf{k}\cdot\mathbf{r}}$; having calculated the integral of which, we have arrived at $\delta(\mathbf{k})$. We must exclude the solution with $\mathbf{k}=0$ from our consideration because $A_\mu \sim e^{ik_\mu x^\mu} \to$ const at $\mathbf{k}=0$, and potentials A_μ in essence transform into U_μ; however, recall that quantities U_μ are equal to zero according to the condition of our problem. Thus, $\mathbf{k}\neq 0$, $J_k^\mu = 0$, and quantity Z yields no contribution.

The situation is different for quantity Y. We write the expression for Y in the form

$$Y = \frac{\hbar}{4\pi}\int \alpha_\mu \frac{a\cdot\tilde{\varepsilon}+\boldsymbol{\alpha}\cdot\tilde{\mathbf{p}}+\beta}{\tilde{\varepsilon}-aE_\ell+\hbar k}\alpha^\mu \frac{d\mathbf{k}}{\tilde{\varepsilon}\cdot k},$$

in which we use, by definition, $\varepsilon = |E|$; note that $\tilde{E} = \boldsymbol{\alpha}\cdot(\mathbf{p}+\hbar\mathbf{k})+\beta$ and, consequently,

$$\tilde{\varepsilon} = \sqrt{(\mathbf{p}+\hbar\mathbf{k})^2 + m^2}.$$

Let us perform the summation with respect to μ. Taking into account that

$$\alpha_\mu \alpha^\mu = -2, \quad \alpha_\mu \beta \alpha^\mu = 4\beta, \quad \text{and} \quad \alpha_\mu(\boldsymbol{\alpha}\cdot\tilde{\mathbf{p}})\alpha^\mu = 2(\boldsymbol{\alpha}\cdot\tilde{\mathbf{p}}),$$

we obtain

$$Y = \frac{\hbar}{2\pi}\int \frac{-a\tilde{\varepsilon}+\boldsymbol{\alpha}\cdot\tilde{\mathbf{p}}+2\beta}{\tilde{\varepsilon}-aE_\ell+\hbar k}\cdot\frac{d\mathbf{k}}{\tilde{\varepsilon}\cdot k} \to \frac{\hbar}{2\pi}\int \frac{(a\hbar k-\varepsilon)+\boldsymbol{\alpha}\cdot\tilde{\mathbf{p}}+2\beta}{\tilde{\varepsilon}-a\varepsilon+\hbar k}\cdot\frac{d\mathbf{k}}{\tilde{\varepsilon}\cdot k} = \hat{Y}.$$

Here, we have replaced $-a\tilde{\varepsilon}$ by $a\hbar k - \varepsilon$. Regarding a, which has two values ± 1, we calculate the average summation with respect to it. With that result, we verify the correctness of the replacement, but with one proviso. One must understand Y as operator $\hat{Y}$ that acts on the vector of state $|\ell\rangle$; as

$$\varepsilon|\ell\rangle = E_\ell|\ell\rangle,$$

the written expressions for Y become substantially equivalent.

Integration needs to be performed. For this purpose, we suggest that the direction of **p** coincides with the z-axis. Such a choice is appropriate because of the fact that the operator of momentum, which is projected on a definite direction, commutes with all variables in the integrand of $\hat{Y}$. In other words, the representation, in which p_z is the diagonal quantity whereas p_x and p_y are equal to zero, is our coordinate system. In what follows, we assume that

$$(\alpha^1, \alpha^2, \alpha^3) = (\alpha_x, \alpha_y, \alpha_z),$$
$$(p^1, p^2, p^3) = (p_x, p_y, p_z),$$
$$(k^1, k^2, k^3) = (k_x, k_y, k_z),$$

etc. So,

$$\hat{Y} = \frac{\hbar}{2\pi} \int \frac{a\hbar k - \varepsilon + \alpha_z(p + \hbar k_z) + 2\beta}{\tilde{\varepsilon} - a\varepsilon + \hbar k} \cdot \frac{d\mathbf{k}}{\tilde{\varepsilon} \cdot k},$$

in which, instead of p_z, we write p. For **k**, we choose cylindrical coordinates; that is,

$$k_x = \rho \cos \phi, \quad k_y = \rho \sin \phi, \quad k_z = k_z,$$
$$k^2 = \rho^2 + k_z^2 \text{ and } d\mathbf{k} = \rho \, d\rho \, dk_z \, d\phi.$$

Instead of ρ, we introduce a new variable of integration,

$$f = \tilde{\varepsilon} + \hbar k = \sqrt{p^2 + 2\hbar p k_z + \hbar^2 k^2 + m^2} + \hbar k;$$

as

$$\left(\frac{df}{d\rho}\right)_{\substack{\phi=\text{const} \\ k_z=\text{const}}} = \frac{\hbar^2 k}{\tilde{\varepsilon}} \cdot \frac{\partial k}{\partial \rho} + \hbar \frac{\partial k}{\partial \rho} = \frac{\hbar \rho f}{k \cdot \tilde{\varepsilon}},$$

$$d\mathbf{k}/k\tilde{\varepsilon} = (df/\hbar f) dk_z \, d\phi.$$

One might promptly perform the integration with respect to ϕ, which yields 2π; consequently,

$$\hat{Y} = \iint \left(\frac{\alpha_z p + 2\beta - \varepsilon}{f - a\varepsilon} + \frac{\hbar \alpha_z}{f - a\varepsilon} k_z + \frac{a\hbar}{f - a\varepsilon} k \right) \frac{df}{f} dk_z.$$

Quantities k and k_z are related to each other through this relation

$$(f - \hbar k)^2 = \varepsilon^2 + 2\hbar p k_z + \hbar^2 k^2,$$

therefore

$$\hbar k = \frac{f^2 - \varepsilon^2}{2f} - \frac{\hbar p}{f} k_z,$$

in which $\varepsilon^2 = p^2 + m^2$. This relation also yields the limits of integration; the limit values of k_z are evidently $\pm k$. Hence, for $-k$ we obtain the lower bound

$$k_{z1} = -\frac{f^2 - \varepsilon^2}{2\hbar(f - p)},$$

and for $+k$ we find the upper bound

$$k_{z2} = \frac{f^2 - \varepsilon^2}{2\hbar(f + p)}.$$

We rewrite $\hat{Y}$ in the form

$$\hat{Y} = \iint \left(\frac{\alpha_z p + 2\beta - \varepsilon}{f - a\varepsilon} + \frac{a(f^2 - \varepsilon^2)}{2f(f - a\varepsilon)} + \left(\frac{\hbar \alpha_z}{f - a\varepsilon} - \frac{a\hbar p}{f(f - a\varepsilon)} \right) k_z \right) \frac{df}{f} dk_z.$$

It is already clear that we should initially integrate with respect to k_z. Here, we encounter two trivial integrals,

$$\int_{k_{z1}}^{k_{z2}} dk_z = \frac{f}{\hbar} \cdot \frac{f^2 - \varepsilon^2}{f^2 - p^2} \quad \text{and} \quad \int_{k_{z1}}^{k_{z2}} k_z dk_z = -\frac{fp}{2\hbar^2} \cdot \frac{(f^2 - \varepsilon^2)^2}{(f^2 - p^2)^2}.$$

Substituting them into $\hat{Y}$, we have

$$\hat{Y} = \frac{1}{\hbar} \int \left(\alpha_z p + 2\beta - \varepsilon + \frac{af}{2} \cdot \frac{f^2 - \varepsilon^2}{f^2 - p^2} - \frac{\alpha_z p}{2} \cdot \frac{f^2 - \varepsilon^2}{f^2 - p^2} \right) \frac{f^2 - \varepsilon^2}{f^2 - p^2} \cdot \frac{df}{f - a\varepsilon}.$$

We represent $f^2 - \varepsilon^2$ in form $(f - a\varepsilon)(f + a\varepsilon)$. Recall, then, that the summation is performed with respect to a, resulting in the disappearance of all terms linear in a and

$$\hat{Y} = \frac{1}{\hbar} \int \left(\frac{\alpha_z p + 2\beta - \varepsilon}{f^2 - p^2} - \frac{\alpha_z p}{2} \cdot \frac{f^2 - \varepsilon^2}{(f^2 - p^2)^2} + \frac{\varepsilon}{2} \cdot \frac{f^2 - \varepsilon^2}{(f^2 - p^2)^2} \right) f \, df.$$

Having further replaced $f^2 - \varepsilon^2$ by $f^2 - p^2 - m^2$, we obtain

$$\hat{Y} = \frac{1}{2\hbar} \int \left(\frac{\alpha_z p + 4\beta - \varepsilon}{f^2 - p^2} - m^2 \cdot \frac{\varepsilon - \alpha_z p}{(f^2 - p^2)^2} \right) f \, df.$$

The latter integral is trivial and needs to be regularized. The method is pertinent here when we restrict the denominator of the total integrand of quantity Y, that is,

$$\tilde{\varepsilon} + \varepsilon + \hbar k < 2\hbar\chi$$

or

$$f < 2\hbar\chi - \varepsilon \approx 2\hbar\chi.$$

In turn, for $k = 0$ quantity f equals ε. Integrating with respect to f in an interval from ε to $2\hbar\chi$ and using $\varepsilon^2 = p^2 + m^2$, we thus have

$$\hat{Y} = \frac{1}{4\hbar}\left[(\alpha_z p + 4\beta - \varepsilon)\ln|f^2 - p^2| + m^2\frac{\varepsilon - \alpha_z p}{f^2 - p^2}\right]\bigg|_{\varepsilon}^{2\hbar\chi}$$

$$\approx \frac{1}{2\hbar}(\alpha_z p + 4\beta - \varepsilon)\ln\left(\frac{2\hbar\chi}{m}\right) - \frac{\varepsilon - \alpha_z p}{4\hbar}.$$

Of course, $\alpha_z p = \boldsymbol{\alpha} \cdot \mathbf{p}$, such that eventually

$$\hat{Y} = \frac{1}{2\hbar}(\boldsymbol{\alpha} \cdot \mathbf{p} + 4\beta - \varepsilon)\ln\left(\frac{2\hbar\chi}{m}\right) - \frac{1}{4\hbar}(\varepsilon - \boldsymbol{\alpha} \cdot \mathbf{p}).$$

Let us act with operator $\hat{Y}$ on the vector of state $|\ell\rangle$; we readily obtain

$$\hat{Y}|\ell\rangle = Y|\ell\rangle.$$

Further, we apply $(\boldsymbol{\alpha} \cdot \mathbf{p} + \beta)|\ell\rangle = E_\ell|\ell\rangle$ and $\varepsilon|\ell\rangle = E_\ell|\ell\rangle$. As a result,

$$Y|\ell\rangle = \frac{q}{\hbar}\beta|\ell\rangle,$$

in which

$$q = \frac{3}{2}\ln\left(\frac{2\hbar\chi}{m}\right) - \frac{1}{4}.$$

For quantity Y, we have essentially derived matrix elements

$$Y_{n\ell} = \frac{q}{\hbar}\langle n|\beta|\ell\rangle,$$

which determine a variation of a creation operator for an electron in state $|\ell\rangle$ in the second order of the perturbation theory. Thus,

$$i\hbar \frac{\partial}{\partial t} \varphi^+_{\ell(2)} = \frac{e^2}{\pi \hbar} q \sum_n \varphi^+_n \langle n|\beta|\ell\rangle e^{i(E_n - E_\ell)t/\hbar}.$$

In these calculations, we have assumed that $c = 1$; here, we can easily restore c, proceeding from the condition of dimension. Therefore,

$$\frac{2\hbar \chi}{m} \to \frac{2\hbar(c\chi)}{mc^2}, \quad \beta = mc\gamma^0$$

and

$$i\hbar \frac{\partial}{\partial t} \varphi^+_{\ell(2)} = \frac{1}{\pi}\left(\frac{e^2}{\hbar c}\right) qmc^2 \sum_n \varphi^+_n \langle n|\gamma^0|\ell\rangle e^{i(E_n - E_\ell)t/\hbar}.$$

In conclusion, we discuss a procedure that is known as the renormalization of mass. We see that $Y_{n\ell} \sim q$, but quantity q is great for large values of χ. How can one avoid this problem? It turns out that one might alter the initial definition of electronic mass; that is, one might use $m \to m + \delta m$, in which m is the observable mass. In the quantized Dirac equation, the additional term with Hamiltonian

$$\delta H = \delta m \cdot c^2 \int \psi^+ \gamma^0 \psi \, d\mathbf{r} = \delta m \cdot c^2 \sum_{nn'} \varphi^+_n \langle n|\gamma^0|n'\rangle \varphi_{n'}$$

thereby arises. According to its conception, δH is supposed to be a quantity of second order; consequently, there appears the additional variation $\varphi^+_{\ell(2)}$, namely

$$i\hbar \frac{\partial}{\partial t} \varphi^+_{\ell(2)} = \left[\delta H', \varphi^+_\ell\right].$$

As we already know, to transfer into the interaction picture we must replace φ^+_n and $\varphi_{n'}$ in δH with $\varphi^{+'}_n = e^{iE_n t/\hbar} \varphi^+_n$ and $\varphi'_{n'} = e^{-iE_{n'} t/\hbar} \varphi_{n'}$, respectively. Thereafter arises the simple commutator

$$\left[\varphi^+_n \varphi_{n'}, \varphi^+_\ell\right],$$

which is equal to $\varphi^+_n \delta_{n'\ell}$. As a result, the additional variation for $\varphi^+_{\ell(2)}$ is

$$i\hbar \frac{\partial}{\partial t} \varphi^+_{\ell(2)} = \delta m \cdot c^2 \sum_n \varphi^+_n \langle n|\gamma^0|\ell\rangle e^{i(E_n - E_\ell)t/\hbar}.$$

The total variation for $\varphi^+_{\ell(2)}$ equals

$$i\hbar \frac{\partial}{\partial t} \varphi^+_{\ell(2)} = \left[\frac{1}{\pi}\left(\frac{e^2}{\hbar c}\right)qm + \delta m\right] c^2 \sum_n \varphi^+_n \langle n|\gamma^0|\ell\rangle e^{i(E_n - E_\ell)t/\hbar}.$$

Using the expression in the square brackets equal to zero, we find

$$\frac{\delta m}{m} = -\frac{q}{\pi}\left(\frac{e^2}{\hbar c}\right)$$

and $i\hbar \partial \varphi^+_{\ell(2)}/\partial t = 0$. The interaction thus has no influence on creation operator φ^+_ℓ of our electron. There is no influence that is at least accurate within the second order with respect to the perturbation.

Physical regularization leads to a necessity of renormalization of electronic mass. These two procedures are inseparable. Moreover, they permit setting the bounds of applicability of the developed quantum electrodynamics. The point is that we cannot choose χ to be too large, because, in this case, we fail to consider δH as a perturbation. At the same time, we cannot choose χ to be too small, because we then fail to correctly regularize the integrals. Let us test these assertions numerically.[13] Suppose that

$$\frac{2\hbar\chi}{mc} = 1000;$$

then,

$$\hbar c \chi = 2.5 \times 10^8 \text{ eV}, \quad \ln\left(\frac{2\hbar\chi}{mc}\right) = 7, \quad \text{and} \quad \frac{\delta m}{m} = \frac{1}{43}.$$

We take into account the value of fine-structure constant $e^2/\hbar c = 1/137$. As we expect, the theory is workable in the range up to several hundred million electron volts; in this case, the additional mass $\delta m = m/43$ is really small. It provides smallness for additional Hamiltonian δH sufficient to apply the latter in calculations according to the perturbation theory. For energy values that transcend the limit approximately 10^9 eV, quantum electrodynamics loses its workability.

Anomalous magnetic moment of the electron

The cogent conclusions regarding spin and a magnetic moment of an electron are the primary triumphs of Dirac's quantum theory of an electron that were confirmed experimentally at the time. Subsequently, it became clear that there exists a small deviation of the magnetic moment from the value equal to one Bohr magneton. According to general considerations, this variation is caused by an interaction of an electron with a quantized field of radiation. To scrutinize the validity of this attribution, our problem, in terms of Y and Z, is to calculate the first nonzero correction to the electronic magnetic moment that, in the literature, is called an anomalous magnetic moment.

What is our frame? A homogeneous magnetic field is directed along the z-axis. In this case, spatial components U_i of a static potential are linear in coordinates, whereas time-like component U_0 is equal to zero. Dirac's operator of energy for an electron has the form

$$E = \alpha \cdot \mathbf{P} + \beta, \quad \text{with } \mathbf{P} = \mathbf{p} + e\mathbf{U}.$$

From Chapter 2, we find that

$$E^2 = \mathbf{P}^2 + m^2 + e\hbar(\boldsymbol{\sigma} \cdot \mathbf{B}),$$

in which $e = |e|$ and $\mathbf{B}$ is the vector of flux density of a homogeneous magnetic field; here, $\boldsymbol{\sigma} \cdot \mathbf{B} = \sigma_z B$. To apply a solution according to the perturbation theory, we suppose that B is sufficiently small and that one might entirely neglect quantity B^2. Hence,

$$|E| \equiv \sqrt{E^2} = \sqrt{\mathbf{P}^2 + m^2} + \frac{e\hbar(\sigma_z B)}{2\sqrt{\mathbf{P}^2 + m^2}}.$$

The expressions for Y and Z are

$$Y = \frac{\hbar}{4\pi} \int \alpha_\mu \Xi \alpha^\mu \frac{d\mathbf{k}}{k} \quad \text{and} \quad Z_{n\ell} = \frac{\hbar}{\pi} \int a J_k^\mu \frac{\langle n|\alpha_\mu e^{-i\mathbf{k}\cdot\mathbf{r}}|\ell\rangle}{E_n - E_\ell + a\hbar k} \frac{d\mathbf{k}}{k},$$

in which

$$\Xi = \frac{a + \tilde{E}/|\tilde{E}|}{|\tilde{E}| - aE_\ell + \hbar k} \quad \text{and} \quad J_k^\mu = \frac{1}{2} \sum_{n'} \langle n'|\alpha^\mu e^{i\mathbf{k}\cdot\mathbf{r}} \eta(-E)|n'\rangle.$$

We first estimate the contribution from quantity Z, if such a contribution exists in principle. We again operate in a representation in which the momentum is diagonal; consequently, $|n\rangle \to |\mathbf{p}\rangle$, summation $\to$ integration and

$$J_k^\mu = \frac{1}{2} \int \langle \mathbf{p}'|\alpha^\mu e^{i\mathbf{k}\cdot\mathbf{r}} \eta(-E)|\mathbf{p}'\rangle \, d\mathbf{p}' = \frac{1}{2} \iint \langle \mathbf{p}'|e^{i\mathbf{k}\cdot\mathbf{r}}|\mathbf{p}''\rangle \langle \mathbf{p}''|\alpha^\mu \eta(-E)|\mathbf{p}'\rangle d\mathbf{p}' \, d\mathbf{p}''.$$

As

$$[e^{i\mathbf{k}\cdot\mathbf{r}}, \mathbf{p}] = i\hbar \frac{\partial e^{i\mathbf{k}\cdot\mathbf{r}}}{\partial \mathbf{r}} = -\hbar \mathbf{k} e^{i\mathbf{k}\cdot\mathbf{r}},$$

then

$$\mathbf{p}e^{i\mathbf{k}\cdot\mathbf{r}} = e^{i\mathbf{k}\cdot\mathbf{r}}(\mathbf{p} + \hbar\mathbf{k}).$$

For this expression, we calculate matrix element $\langle\mathbf{p}'|\ldots|\mathbf{p}''\rangle$; as a result,

$$(\mathbf{p}' - \mathbf{p}'' - \hbar\mathbf{k})\langle\mathbf{p}'|e^{i\mathbf{k}\cdot\mathbf{r}}|\mathbf{p}''\rangle = 0,$$

therefore

$$\langle\mathbf{p}'|e^{i\mathbf{k}\cdot\mathbf{r}}|\mathbf{p}''\rangle = \delta(\mathbf{p}' - \mathbf{p}'' - \hbar\mathbf{k}).$$

Thus,

$$J_k^\mu = \frac{1}{2}\int \langle\mathbf{p}''|\alpha^\mu \eta(-E)|\mathbf{p}'' + \hbar\mathbf{k}\rangle\, d\mathbf{p}''.$$

Further,

$$\eta(-E) = \frac{1 - E/|E|}{2};$$

that is, η represents a function of energy, which is equal to $\boldsymbol{\alpha}\cdot(\mathbf{p}+e\mathbf{U})+\beta$. Potential $\mathbf{U}$ is linear with respect to the spatial coordinates. In the momentum representation, we encounter the matrix elements of type $\langle\mathbf{p}'|x|\mathbf{p}''\rangle$. Taking into account that $[x, p_x] = i\hbar$, we have

$$\langle\mathbf{p}'|x|\mathbf{p}''\rangle = i\hbar\frac{\delta(\mathbf{p}'' - \mathbf{p}')}{p_x'' - p_x'}.$$

We see that the matrix elements that differ from zero are $\langle\mathbf{p}'|E|\mathbf{p}''\rangle$ that are only infinitesimally distant from the diagonal elements. Thus, $J_k^\mu = 0$ for $\mathbf{k} \neq 0$ and, like the case in which a static field is absent, quantity Z yields no contribution.

We proceed to consider the contribution from Y. For this purpose, we introduce an auxiliary quantity

$$\hat{Y}_0 = \frac{\hbar}{4\pi}\int \alpha_\mu \Xi_0 \alpha^\mu \frac{d\mathbf{k}}{k},$$

in which

$$\Xi_0 = \frac{a + (\boldsymbol{\alpha}\cdot\tilde{\mathbf{P}} + \beta)(\tilde{\mathbf{P}}^2 + m^2)^{-1/2}}{(\tilde{\mathbf{P}}^2 + m^2)^{1/2} - a(\mathbf{P}^2 + m^2)^{1/2} + \hbar k}.$$

Operator $\hat{Y}_0$ corresponds to the case in which in $|E|$ we neglect the quantity that is proportional to field B. Note that $\hat{Y}_0$ has practically the same form as quantity $\hat{Y}$ for the case of a free electron that we consider in the preceding section. For $\hat{Y}_0$, one might apply a solution previously obtained, with $\mathbf{p}$ preliminarily replaced with $\mathbf{P}$ and ε replaced with $(\mathbf{P}^2+m^2)^{1/2}$; that is,

$$\hat{Y}_0 \to \frac{1}{2\hbar}(\boldsymbol{\alpha}\cdot\mathbf{P}+4\beta-(\mathbf{P}^2+m^2)^{1/2})\ln\left(\frac{2\hbar\chi}{m}\right) - \frac{1}{4\hbar}((\mathbf{P}^2+m^2)^{1/2}-\boldsymbol{\alpha}\cdot\mathbf{P}).$$

However, quantity $\mathbf{P}$ is a formal momentum that fails to commute with itself. To progress, one must make a reasonable supposition that, like $\mathbf{p}$, "momentum" $\mathbf{P}$ is a small quantity. We represent Ξ_0 in the form of an expansion with respect to $\mathbf{P}$ up to terms of second order that contain $\mathbf{P}^2$, or $(\mathbf{k}\cdot\mathbf{P})^2$, or $(\boldsymbol{\alpha}\cdot\mathbf{P})(\mathbf{k}\cdot\mathbf{P})$, and neglect the cubic terms of order $|\mathbf{P}|^3$. After integration with respect to all possible directions of $\boldsymbol{l}=\mathbf{k}/k$, there remain the terms of type $\mathbf{P}^2$. The terms containing $\boldsymbol{l}$ in the first power yield no contribution. Thus, we apply this written expression for $\hat{Y}_0$ that is accurate only within $\mathbf{P}^2$, when $\mathbf{P}$ behaves like a commuting variable. Using $(\mathbf{P}^2+m^2)^{1/2}\approx m+\mathbf{P}^2/2m$, we have

$$\hat{Y}_0 = \frac{1}{2\hbar}\left[\boldsymbol{\alpha}\cdot\mathbf{P}+\beta-\left(m+\frac{\mathbf{P}^2}{2m}\right)\right]\left(\ln\left(\frac{2\hbar\chi}{m}\right)+\frac{1}{2}\right)+q\beta,$$

in which we explicitly isolate the part that contains $q=(3/2)\ln(2\hbar\chi/m)-1/4$; subsequently, one might eliminate it through a renormalization of mass.

Let us write the expression for Ξ, for which we have

$$\Xi = \frac{a+(\boldsymbol{\alpha}\cdot\tilde{\mathbf{P}}+\beta)(\tilde{\mathbf{P}}^2+m^2+e\hbar(\sigma_z B))^{-1/2}}{(\tilde{\mathbf{P}}^2+m^2+e\hbar(\sigma_z B))^{1/2}-aE_\ell+\hbar k}.$$

We see that the distinction between Ξ and Ξ_0 is at least of order B and δ, in which $\delta = E_\ell - \sqrt{\mathbf{P}^2+m^2}$. Using

$$B=0 \text{ and } E_\ell - \sqrt{\mathbf{P}^2+m^2}\approx 0,$$

we directly obtain that Ξ and Ξ_0 are equal to each other. Quantity δ equals zero in a weak sense because

$$E_\ell - \sqrt{\mathbf{P}^2+m^2} \approx |E| - \sqrt{\mathbf{P}^2+m^2} \sim B.$$

Further, one might expand $\Xi - \Xi_0$ in a series with respect to B and δ, from which, restricting oneself to the first order, one might obtain the expression that determines the energy variation of an electron in a magnetic field.

Momentarily overlooking that we work with operators, let us calculate the expansion. For the numerator, we thus have

$$a + (\boldsymbol{\alpha} \cdot \tilde{\mathbf{P}} + \beta)(\tilde{\mathbf{P}}^2 + m^2 + e\hbar(\sigma_z B))^{-1/2}$$

$$\approx a + (\boldsymbol{\alpha} \cdot \tilde{\mathbf{P}} + \beta)\left[(\tilde{\mathbf{P}}^2 + m^2)^{-1/2} - \frac{e\hbar(\sigma_z B)}{2(\tilde{\mathbf{P}}^2 + m^2)^{3/2}}\right].$$

Assuming $X = \sqrt{\tilde{\mathbf{P}}^2 + m^2} - a(\mathbf{P}^2 + m^2) + \hbar k$, we expand the denominator; evidently,

$$\left(\sqrt{\tilde{\mathbf{P}}^2 + m^2 + e\hbar(\sigma_z B)} - a\delta - a\sqrt{\mathbf{P}^2 + m^2} + \hbar k\right)^{-1} \approx \frac{1}{X} - \frac{e\hbar(\sigma_z B)}{2X^2\sqrt{\tilde{\mathbf{P}}^2 + m^2}} + a\frac{\delta}{X^2}.$$

Recalling that Ξ is converted into Ξ_0 at $B = 0$ and $\delta = 0$, for difference $\Xi - \Xi_0$, neglecting the terms of order $\mathbf{BP}$, we obtain the expression

$$-\frac{e\hbar(\sigma_z B)\beta}{2X(\tilde{\mathbf{P}}^2 + m^2)^{3/2}} - \frac{1}{X^2}\left(a + \frac{\beta}{\sqrt{\tilde{\mathbf{P}}^2 + m^2}}\right)\left(\frac{e\hbar(\sigma_z B)}{2\sqrt{\tilde{\mathbf{P}}^2 + m^2}} - a\delta\right).$$

To proceed to the operators, one must define the order of the factors obtained here for the following reason. Momentum $\mathbf{P}$ is small, such that for the first nonzero approximation we can use $\mathbf{P} = 0$; then $\tilde{\mathbf{P}} = \hbar \mathbf{k}$, $\delta = E_\ell - m$, and

$$\Xi - \Xi_0 = -\frac{\beta e\hbar(\sigma_z B)}{2g^3(g - am + \hbar k)} - \left(a + \frac{\beta}{g}\right)\frac{(e\hbar(\sigma_z B)/2g - a(E_\ell - m))}{(g - am + \hbar k)^2},$$

in which $g = \sqrt{\hbar^2 k^2 + m^2}$, and the initial order of factors has been maintained only for β and σ_z.

Performing the summation with respect to μ, we find the expression for $\alpha_\mu(\Xi - \Xi_0)\alpha^\mu$. Directly, $(\alpha^\mu)^2 = 1$, $\alpha_\mu \alpha^\mu = -2$ and $\alpha_\mu \beta \alpha^\mu = 4\beta$; also,

$$\alpha_\mu \sigma_z \alpha^\mu = (\alpha^0)^2 \sigma_z + \alpha_i \sigma_z \alpha^i = \sigma_z - \sum_i \rho_1^2 \sigma^i \sigma_z \sigma^i = 2\sigma_z$$

and

$$\alpha_\mu \beta \sigma_z \alpha^\mu = (\alpha^0)^2 \beta \sigma_z - \beta \alpha_i \sigma_z \alpha^i = 0.$$

As a result,

$$\alpha_\mu(\Xi - \Xi_0)\alpha^\mu = -\frac{1}{(g - am + \hbar k)^2}\left(a\frac{e\hbar(\sigma_z B)}{g} + 2a^2(E_\ell - m) - \frac{4a\beta}{g}(E_\ell - m)\right),$$

in which one might replace a^2 by unity.

According to the chosen approximation,

$$|E| = \sqrt{\mathbf{P}^2 + m^2} + \frac{e\hbar(\sigma_z B)}{2\sqrt{\mathbf{P}^2 + m^2}};$$

because quantity $|\mathbf{P}|$ is small, we use

$$E_\ell - m \approx \frac{e\hbar(\sigma_z B)}{2m}.$$

This sense is weak but sufficient for us, because we must know expression $\hat{Y} - \hat{Y}_0$ only so far as we know $(\hat{Y} - \hat{Y}_0)|\ell\rangle$. Thus,

$$\alpha_\mu(\Xi - \Xi_0)\alpha^\mu \approx -\frac{e\hbar(\sigma_z B)}{(g - am + \hbar k)^2}\left(\frac{a}{g} + \frac{1}{m} - \frac{2a}{g}\right) = -\frac{e\hbar(\sigma_z B)}{(g - am + \hbar k)^2}\left(\frac{g - am}{gm}\right),$$

in which we take into account that $\beta - m \approx 0$; this relation follows directly from weak equality $E_\ell \approx \boldsymbol{\alpha} \cdot \mathbf{P} + \beta$, which one might rewrite in the form

$$[(E_\ell - m) - \boldsymbol{\alpha} \cdot \mathbf{P}] - (\beta - m) \approx 0,$$

in which the quantities between square brackets are small and might be omitted.

To lead the latter expression for $\alpha_\mu(\Xi - \Xi_0)\alpha^\mu$ to a form more convenient for integration, we write

$$(g - am + \hbar k)(g + am - \hbar k) = 2a\hbar km,$$

hence

$$(g - am + \hbar k)^{-2} = \frac{(g + am - \hbar k)^2}{4\hbar^2 k^2 m^2} = \frac{(g - \hbar k)(g + am)}{2\hbar^2 k^2 m^2}.$$

Further, taking into account that $(g + am)(g - am) = g^2 - m^2 = \hbar^2 k^2$, we have

$$\alpha_\mu(\Xi - \Xi_0)\alpha^\mu \approx -e\hbar(\sigma_z B) \cdot \frac{g - \hbar k}{2 m^3 g}.$$

Using spherical coordinates, we readily perform the integration with respect to $\mathbf{k}$. For this purpose, we recall that $\hat{Y} - \hat{Y}_0 \approx Y - Y_0$; then,

$$Y - Y_0 = -\frac{e\hbar^2(\sigma_z B)}{4\pi}\int \frac{(g - \hbar k)}{2m^3 g} \cdot \frac{4\pi k^2 dk}{k}$$

$$= -\frac{e(\sigma_z B)}{2m^3}\int_0^{\hbar X}\left(1 - \frac{\hbar k}{\sqrt{m^2 + \hbar^2 k^2}}\right)(\hbar k) \cdot d(\hbar k),$$

in which we apply regularization in the form of a simple truncation of the region of too large values of k.

Radiative corrections

The integral is calculated in a trivial manner, namely,

$$\int \left(1 - \frac{x}{\sqrt{m^2 + x^2}}\right) x\, dx = \frac{x^2}{2} - \frac{x\sqrt{m^2 + x^2}}{2} + \frac{m^2}{2}\ln\left|x + \sqrt{m^2 + x^2}\right|.$$

Assuming

$$(\hbar\chi)^2 - (\hbar\chi)\sqrt{m^2 + \hbar^2\chi^2} = -\frac{m^2}{2}$$

and, in the logarithm,

$$\sqrt{m^2 + \hbar^2\chi^2} = \hbar\chi,$$

we obtain

$$Y - Y_0 = -\frac{e(\sigma_z B)}{4m}\left(\ln\left(\frac{2\hbar\chi}{m}\right) - \frac{1}{2}\right).$$

There remains to recall the expression for Y_0, namely $\hat{Y}_0|\ell\rangle = Y_0|\ell\rangle$, and $\boldsymbol{\alpha}\cdot\mathbf{P} + \beta \to E_\ell$; we omit $\mathbf{P}^2$ because of the smallness and we omit $q\beta$ because of the renormalization of mass. As a result,

$$Y_0 = \frac{E_\ell - m}{2\hbar}\left(\ln\left(\frac{2\hbar\chi}{m}\right) + \frac{1}{2}\right).$$

Finally, $E_\ell - m \approx e\hbar(\sigma_z B)/2m$; consequently,

$$Y|\ell\rangle = \frac{e(\sigma_z B)}{4m}|\ell\rangle.$$

The result is simple and remarkable. First, it shows that the nondiagonal matrix elements of quantity Y are equal to zero; hence, the eigenfunctions of an electron remain unperturbed. Second, an interaction of an electron with a field of radiation is characterized by the appearance of additional energy $\Delta E_B = (e^2/\pi)Y_{\ell\ell}$; obviously,

$$\Delta E_B = \frac{e^3}{2\pi\hbar m}(\mathbf{s}\cdot\mathbf{B}),$$

in which $\mathbf{s} = \hbar\boldsymbol{\sigma}/2$ is the spin of an electron. Using m equal to mc^2, we restore c; as a result,

$$\Delta E_B = \frac{1}{2\pi}\left(\frac{e^2}{\hbar c}\right)(\boldsymbol{\mu}_B\cdot\mathbf{B}).$$

This result, obtained in a frame of Dirac's theory, agrees satisfactorily with the experiment. An electron thus possesses an additional magnetic moment that, in Bohr magnetons, is equal to fine-structure constant $e^2/\hbar c$ divided by 2π.

On the history of radiative corrections

The anomalous magnetic moment of an electron and the Lamb shift of atomic levels of hydrogen have played a key role in the development of contemporary quantum electrodynamics. The theoretical investigations were, however, far from immediately able to confirm the unique experimental facts at that moment of time. For instance, the calculation of the self-energy of an electron, as in classical electrodynamics, yielded an infinite value. The situation was analogous to the calculation of the electromagnetic shift of atomic levels. As Bethe said, "one might overlook the Lamb shift, because the latter was infinite in all the then existing theories." This scenario was generally applicable to all infinities, which were simply discarded. Do the infinities lack a physical meaning?

It subsequently became clear that the reply to the appearance of infinities was hidden within them. Let us recall that on quantizing an electromagnetic field, we have discarded an infinite c-number, associating it with the zero-point vibrations of the field or, as one says, with an electromagnetic vacuum. One might ignore the vacuum, but only when we consider the transitions of an electron between excited states. Moreover, there exists an electron–positron vacuum that is called the Dirac sea. To calculate the radiative corrections, we must take into account the interaction of electrons with the virtual electron–positron pairs of the Dirac sea, and not only with the zero-point vibrations of an electromagnetic field. As a result of the interaction with the "vacuum," remarkable effects of quantum electrodynamics arise — the anomalous magnetic moment and the Lamb shift of atomic levels.

We have already obtained the value for an additional magnetic moment. In 1947, Nafe, Nelson, and Rabi, measuring the hyperfine structure of hydrogen and deuterium, discovered it experimentally. One year later, Kusch and Foley repeated the radiospectroscopic measurements of the Zeeman effect for sodium and gallium. It was convincingly proven that, for an electron, the ratio of the magnetic moment to the spin differs in value $g(e/2mc)$, in which $g = 2$. Breit heuristically suggested that $g = 2(1 + e^2/2\pi\hbar c)$. In 1948, Schwinger, having developed the powerful formalism of canonical quantization with the elimination of infinities, calculated g and, thus, proved Breit's supposition. In the same year, Luttinger, having applied an ordinary perturbation theory, showed that if, when calculating the energy of an electron in a homogeneous magnetic field, attention is focused only on the corrections that contain a magnetic field vector, then one might also obtain the correct value for the anomalous magnetic moment without a special formalism to eliminate infinities. In 1950, Karplus and Kroll calculated the next correction to the magnetic moment of an electron up to the fourth order that is accurate within $(e^2/\hbar c)^2$.

Seven years later, Sommerfield and Petermann refined this correction; as a result, in Bohr magnetons, it amounted to

$$\left(\frac{e^2}{\pi\hbar c}\right)^2 \left(\frac{197}{144} + \frac{\pi^2}{12} - \frac{\pi^2}{2}\ln 2 + \frac{3}{4}\zeta(3)\right) = -0.328\left(\frac{e^2}{\pi\hbar c}\right)^2$$

that also agrees satisfactorily with the experiment.

Let us review the history of the electromagnetic Lamb shift. During the last 30 years of the twentieth century, Houston and Williams first performed, although unconvincingly, the spectroscopic measurements of the shift of level $2S_{1/2}$ relative to level $2P_{1/2}$ for hydrogen. According to Dirac's theory, these levels must coincide exactly. In 1947, applying more precise methods of radiospectroscopy, Lamb and Retherford showed that this shift really exists and equals approximately 1000 MHz. It became clear that there is no degeneracy; level $2S_{1/2}$ lies slightly above level $2P_{1/2}$. The transition of an electron from state $2S_{1/2}$ to state $1S_{1/2}$ is characterized by a small probability because of the forbidden state of dipolar and quadrupolar emission mechanisms. It is more practical to proceed such that an electron first transits to state $2P_{1/2}$, and only then to state $1S_{1/2}$; Lamb and Retherford used this circumstance in their experiments. A beam of hydrogen atoms, which were in states $2S_{1/2}$ and $2P_{1/2}$, was incident on a metallic target. The atoms in metastable state $2S_{1/2}$ contributed an emission of electrons from a metal so that a current appeared. If, in addition, microwave electromagnetic radiation was imposed on the beam, then the current vanished at frequency $\nu \sim 10$ GHz. This frequency was properly recognized as resonant for the transition $2S_{1/2} \to 2P_{1/2}$. Subsequently, the atoms were almost instantaneously passed into state $1S_{1/2}$. Such is the essence of their experiment. In the same year, Bethe promulgated his work regarding an electromagnetic shift of energy levels. Referring to the result of Pasternack that the Lamb shift is independent of a nuclear interaction and the result of Uehling indicating that there is only a small influence of a polarization of Dirac's electron–positron vacuum on the shift, he calculated the necessary effect in a nonrelativistic approximation in excellent agreement with the experimental value. Bethe's formula has played an important role in the development of quantum electrodynamics "without infinities." We consider its derivation in detail.[15]

Bethe's formula

In a framework of an ordinary perturbation theory, we express the second correction to energy E_ℓ as

$$W_\ell = \sum_{n \neq \ell} \frac{1}{E_\ell - E_n} |\langle n|\text{perturbation}|\ell\rangle|^2.$$

As a perturbation, we have $H_{\text{int}} = e(\alpha \cdot \mathbf{A}) \to e(\mathbf{v} \cdot \mathbf{A})/c$, in which the latter conversion corresponds to the nonrelativistic consideration. The summation is performed with respect to all intermediate states of a system comprising an atomic electron and a photon; that is,

$$E_n \to E_n + \hbar\omega, |n\rangle \to |n;1\rangle = |n\rangle \cdot |\mathbf{k}, \lambda\rangle \text{ and } \sum_n \to \sum_n \sum_{\mathbf{k},\lambda}.$$

Here, E_n and $|n\rangle$ represent, respectively, the energy and state vector of an atomic electron; $|\mathbf{k}, \lambda\rangle$ or $|1\rangle$ is the state vector of an intermediate photon with wave vector $\mathbf{k}$ and index of polarization λ,

$$\omega = c|\mathbf{k}|.$$

In an initial state, there exists one electron in state $|\ell\rangle$ and there is no photon; hence, $|\ell\rangle \to |\ell;0\rangle$. Consequently,

$$W_\ell = \frac{e^2}{c^2} \sum_n \sum_{\mathbf{k},\lambda} \frac{|\langle n;1|\mathbf{v}\cdot\mathbf{A}|\ell;0\rangle|^2}{E_\ell - E_n - \hbar\omega}.$$

According to the best traditions of electrodynamics, we further expand $\mathbf{A}$ with respect to plane waves,

$$\mathbf{A} = \sum_{\mathbf{k}'\lambda'}{}' (q_{\mathbf{k}'\lambda'}\mathbf{A}_{\mathbf{k}'\lambda'} + q^+_{\mathbf{k}'\lambda'}\mathbf{A}^*_{\mathbf{k}'\lambda'}), \quad \mathbf{A}_{\mathbf{k}'\lambda'} = \sqrt{\frac{4\pi c^2}{\tau}}\mathbf{e}_{\lambda'}e^{i\mathbf{k}'\cdot\mathbf{r}},$$

in which, for the indices of summation, the primes are introduced to distinguish from the corresponding values in W_ℓ; the prime on the summation symbol emphasizes that the summation is performed over the hemisphere of directions of $\mathbf{k}'$. Moreover, $\mathbf{e}_\lambda$ are the unit real-valued vectors of polarization,

$$\mathbf{k} \cdot \mathbf{e}_\lambda = 0 \text{ and } \mathbf{e}_\lambda \cdot \mathbf{e}_{\lambda'} = \delta_{\lambda\lambda'},$$

in which $\lambda = 1$, 2, and 3. Quantity $\sqrt{4\pi c^2/\tau}$ is the normalization factor with respect to a volume τ of quantization. Quantities q and q^+ are typical Bose operators for destruction and creation of photons, for which a commutator is equal to the quantum mechanical amplitude $\sqrt{\hbar/2\omega}$.

Substituting $\mathbf{A}$ into the expression for W_ℓ, we see that

$$q_{\mathbf{k}'\lambda'}|0\rangle = 0 \text{ and } \langle 1|q^+_{\mathbf{k}'\lambda'}|0\rangle = \sqrt{\frac{\hbar}{2\omega}}\delta_{\mathbf{k}'\mathbf{k}}\delta_{\lambda'\lambda},$$

yielding

$$W_\ell = \frac{\hbar e^2}{2c^2} \sum_n \sum_{\mathbf{k},\lambda} \frac{1}{\omega} \frac{|\langle n|\mathbf{v}\cdot\mathbf{A}^*_{\mathbf{k}\lambda}|\ell\rangle|^2}{E_\ell - E_n - \hbar\omega}.$$

Applying the known rule $\sum_{\mathbf{k}} \to (\tau/(2\pi)^3) \int d\mathbf{k}$, we proceed to the integral from the sum with respect to $\mathbf{k}$,

$$W_\ell = \frac{\hbar e^2}{(2\pi)^2} \sum_{n,\lambda} \int \frac{d\mathbf{k}}{\omega} \cdot \frac{|\langle n|\mathbf{v}\cdot\mathbf{e}_\lambda|\ell\rangle|^2}{E_\ell - E_n - \hbar\omega},$$

in which, according to a dipole approximation, we use $e^{-i\mathbf{k}\cdot\mathbf{r}} \approx 1$. In spherical coordinates, $d\mathbf{k} = k^2 \sin\theta\, dk\, d\theta\, d\phi$; as $\mathbf{k}\cdot\mathbf{e}_\lambda = 0$, one might replace $\mathbf{v}\cdot\mathbf{e}_\lambda$ by $v\sin\theta$, in which θ is the angle between vectors $\mathbf{k}$ and $\mathbf{v}$. Further, one might perform the integration over the angles to yield

$$\int_0^\pi \sin^3\theta\, d\theta \int_0^{2\pi} d\phi = \frac{8\pi}{3}.$$

Thus,

$$W_\ell = \frac{2e^2}{3\pi\hbar c^3} \int G\, dG \sum_n \frac{|\mathbf{v}_{n\ell}|^2}{E_\ell - E_n - G},$$

in which $G = \hbar\omega$. Bethe's work opens with this formula.

Supposing that the relativistic theory must lead to a natural truncation for the integral, Bethe spreads the integration from zero to some maximum value of the energy of a photon, setting this value equal to mc^2; then,

$$W_\ell = -\frac{2e^2}{3\pi\hbar c^3} \sum_n \int_0^{mc^2} |\mathbf{v}_{n\ell}|^2 dG + \frac{2e^2}{3\pi\hbar c^3} \sum_n \int_0^{mc^2} \frac{|\mathbf{v}_{n\ell}|^2(E_n - E_\ell)}{E_n - E_\ell + G} dG.$$

The first expression here represents the variation of kinetic energy of an electron,

$$-\frac{4e^2}{3\pi\hbar c} \cdot \frac{m(\mathbf{v}^2)_{\ell\ell}}{2};$$

having included an additional electromagnetic mass of order $e^2/\hbar c$ into the mass that is observable in the experiment, one might eliminate it with the aid of a renormalization of mass. The second expression can be easily integrated; this expression, which we denote as W'_ℓ, properly defines the Lamb shift. Replacing a velocity by a momentum, we have

$$W'_\ell = \frac{2e^2}{3\pi\hbar m^2 c^3} \sum_n |\mathbf{p}_{n\ell}|^2 (E_n - E_\ell) \ln\left(\frac{mc^2}{|E_n - E_\ell|}\right),$$

in which we neglect all differences $E_n - E_\ell$ relative to upper bound mc^2. Moreover, because quantity mc^2 is great, one might remove the logarithm from under the summation sign, assuming that

$$\ln\left(\frac{mc^2}{|E_n - E_\ell|}\right) \to \ln\left(\frac{mc^2}{\langle E_n - E_\ell\rangle}\right),$$

in which $\langle E_n - E_\ell\rangle$ designates the average value of $|E_n - E_\ell|$. The sum $\sum_n |\mathbf{p}_{n\ell}|^2 (E_n - E_\ell)$ remains to be calculated. Obviously,

$$\sum_n |\mathbf{p}_{n\ell}|^2 (E_n - E_\ell) = \frac{1}{2}\langle\ell|[[\mathbf{p}, H], \mathbf{p}]|\ell\rangle;$$

however,

$$[\mathbf{p}, H] = -i\hbar \nabla V \quad \text{and} \quad [[\mathbf{p}, H], \mathbf{p}] = \hbar^2 \nabla^2 V,$$

in which V is the electrostatic potential energy of an atomic electron and, hence,

$$W'_\ell = \frac{\hbar e^2}{3\pi m^2 c^3} \langle\ell|\nabla^2 V|\ell\rangle \ln\left(\frac{mc^2}{\langle E_n - E_\ell\rangle}\right).$$

For an atom of hydrogen type with nuclear charge Ze, we have

$$V(r) = -\frac{Ze^2}{r}.$$

Further, $\nabla^2(1/r) = -4\pi\delta(r)$; consequently,

$$\langle\ell|\nabla^2 V|\ell\rangle = \int |\langle\mathbf{r}|\ell\rangle|^2 \nabla^2 V \, d\mathbf{r} = 4\pi Ze^2 |\langle 0|\ell\rangle|^2.$$

For an electron with nonzero orbital angular momentum, wave function $\langle\mathbf{r}|\ell\rangle$ vanishes at $\mathbf{r} = 0$. However, if an electron occupies an S-state, then

$$\langle 0|\ell\rangle = \sqrt{\frac{Z^3}{\pi\nu^3 a^3}},$$

in which $a = \hbar^2/me^2$ and $\nu = 1, 2, \ldots$; ν is the principal quantum number. Thus, eventually,

$$W'_\ell = \frac{8}{3\pi}\left(\frac{e^2}{\hbar c}\right)^3 \cdot \frac{m(eZ)^4}{2\hbar^2 \nu^3} \ln\left(\frac{mc^2}{\langle E_n - E_\ell\rangle}\right).$$

This is the famous Bethe formula.

This formula has been primarily applied to the atom of hydrogen. Bethe estimated numerically the average value of the difference $\langle E_n - E_\ell \rangle$ of energies and evaluated the logarithm, which became equal to 7.63; as a result, for hydrogen in an S-state,

$$W'_\ell = 1040 \text{ MHz}.$$

For a P-state, one might neglect the shift. We see that the agreement with the experiment is excellent. An experiment subsequently refined the value of the Lamb shift to 1057.8 MHz.

One astonishing conclusion of that work deserves mention. Together with the necessity of the subtraction of an infinite electromagnetic mass and with the procedure of regularization of the integral, Bethe noted that the shift of the atomic level, which is caused by the interaction with the field of radiation, is the real effect and has a finite value. One might have suspected a mistrust of quantum electrodynamics at that time. Moreover, this scenario has not changed. Quantum electrodynamics fails to work without regularizations and renormalizations.

Bethe indicated further that the shift grows less rapidly than Z^4 because of the variation of $\langle E_n - E_\ell \rangle$ when transiting to another atom. For the shift of the $2S$-level of a helium atom, he obtained a value that is 13 times that of the corresponding value for hydrogen.

Electromagnetic shift of atomic levels

Let us return to our solution for quantity Y to define the Lamb shift. Relative to Y, quantity Z, as seen further in the next section, yields a smaller contribution to an electromagnetic shift; moreover, this contribution has an opposite sign. So,

$$Y = \frac{\hbar}{4\pi} \int \alpha_\mu \Xi \alpha^\mu \frac{d\mathbf{k}}{k},$$

in which

$$\Xi = \frac{a + \tilde{E}/|\tilde{E}|}{|\tilde{E}| - aE_\ell + \hbar k};$$

the operator of energy is given by the expression

$$E = \boldsymbol{\alpha} \cdot \mathbf{p} + \beta - eU_0.$$

We have electrostatic potential $U_0(\mathbf{r})$, which we denote as $U(\mathbf{r})$. This notation is appropriate because magnetic vector potential $\mathbf{U}$ is absent. If $\mathbf{U} = 0$, then

$$Y_0 = \frac{\hbar}{4\pi} \int \alpha_\mu \Xi_0 \alpha^\mu \frac{d\mathbf{k}}{k},$$

in which

$$\Xi_0 = \frac{a + \tilde{\xi}/\tilde{\varepsilon}}{\tilde{\varepsilon} - a\varepsilon + \hbar k},$$

$$\xi = \boldsymbol{\alpha} \cdot \mathbf{p} + \beta \text{ and } \varepsilon = |\xi| = (\mathbf{p}^2 + m^2)^{1/2}.$$

To estimate the sought shift, we must focus only on difference $Y - Y_0$, which yields the variation of an electronic energy that arises from an interaction with radiation when only a static electric field exists. Moreover, together with momentum $\mathbf{p}$ of an electron, quantity U is supposed to be small. Neglecting U^2, we write

$$E^2 = \varepsilon^2 - e(\xi U + U\xi),$$

therefore

$$|E| = \varepsilon - \frac{1}{2}(\xi F + F\xi),$$

in which, to find quantity F, one must square $|E|$ and compare the result with E^2. With an accuracy up to terms linear in F, we have

$$|E|^2 = \varepsilon^2 - \frac{1}{2}\xi(\varepsilon F + F\varepsilon) - \frac{1}{2}(\varepsilon F + F\varepsilon)\xi;$$

consequently,

$$\varepsilon F + F\varepsilon = 2eU.$$

Having calculated the matrix element $\langle \mathbf{p}'|\ldots|\mathbf{p}''\rangle$ for the obtained relation, it is convenient to proceed to the momentum representation. As a result,

$$\langle \mathbf{p}'|F|\mathbf{p}''\rangle = 2e\frac{\langle \mathbf{p}'|U|\mathbf{p}''\rangle}{\varepsilon' + \varepsilon''},$$

in which $\varepsilon' = (\mathbf{p}'^2 + m^2)^{1/2}$ and $\varepsilon'' = (\mathbf{p}''^2 + m^2)^{1/2}$.

If, in addition, we expand U into a Fourier integral

$$U(\mathbf{r}) = \int U_k e^{-i\mathbf{k}\cdot\mathbf{r}}\, d\mathbf{k}$$

and recall that $\langle \mathbf{p}'|e^{-i\mathbf{k}\cdot\mathbf{r}}|\mathbf{p}''\rangle = \delta(\mathbf{p}' - \mathbf{p}'' + \hbar\mathbf{k})$, then one might lead matrix F to the form

$$\langle \mathbf{p}'|F|\mathbf{p}''\rangle = \frac{2e}{\hbar^3(\varepsilon' + \varepsilon'')} U_{(\mathbf{p}''-\mathbf{p}')/\hbar}.$$

Radiative corrections

To approximate the expression for $E/|E|$, we apply an expansion

$$\frac{1}{T+\Delta} = \frac{1}{T} - \frac{1}{T}\Delta\frac{1}{T} + \cdots,$$

which is valid for some operators T and Δ under the condition that quantity Δ is small. Then,

$$\frac{1}{|E|} = \frac{1}{\varepsilon} + \frac{1}{\varepsilon} \cdot \frac{(\xi F + F\xi)}{2} \cdot \frac{1}{\varepsilon}$$

and, with an accuracy up to the terms of first order of smallness,

$$\frac{E}{|E|} = \frac{\xi - eU}{\varepsilon} + \frac{1}{\varepsilon} \cdot \frac{(\xi^2 F + \xi F\xi)}{2} \cdot \frac{1}{\varepsilon} = \frac{\xi}{\varepsilon} - \frac{1}{\varepsilon}K\frac{1}{\varepsilon}, \quad K = \frac{1}{2}(\varepsilon F\varepsilon - \xi F\xi),$$

in which we use the relation $\varepsilon F + F\varepsilon = 2eU$ and the fact that $\xi^2 = \varepsilon^2$.

To estimate the order of quantity K, we consider it in detail. So, as

$$\xi F\xi = (\boldsymbol{\alpha} \cdot \mathbf{p})F(\boldsymbol{\alpha} \cdot \mathbf{p}) + \beta F(\boldsymbol{\alpha} \cdot \mathbf{p}) + (\boldsymbol{\alpha} \cdot \mathbf{p})F\beta + \beta^2 F,$$

for the matrix element of K in the momentum representation, we obtain

$$\langle \mathbf{p}'|K|\mathbf{p}''\rangle = \frac{1}{2}\langle \mathbf{p}'|F|\mathbf{p}''\rangle(\varepsilon'\varepsilon'' - (\boldsymbol{\alpha} \cdot \mathbf{p}')(\boldsymbol{\alpha} \cdot \mathbf{p}'') - \beta\boldsymbol{\alpha} \cdot (\mathbf{p}'' - \mathbf{p}') - m^2).$$

Here,

$$\varepsilon'\varepsilon'' \sim m^2 + o(\mathbf{p}^2);$$

expression $\beta\boldsymbol{\alpha} \cdot (\mathbf{p}'' - \mathbf{p}')$ vanishes through the equality

$$\alpha_\mu(\beta\boldsymbol{\alpha})\alpha^\mu = \beta(\boldsymbol{\alpha} - \alpha_i\boldsymbol{\alpha}\alpha^i) = 0,$$

in which the summation is performed with respect to μ. Quantity $\alpha_\mu\langle\mathbf{p}'|K|\mathbf{p}''\rangle\alpha^\mu$ is thus small and becomes, at most, of order $U\mathbf{p}^2$. Together with this quantity, the expression

$$\alpha_\mu\left(\frac{E}{|E|} - \frac{\xi}{\varepsilon}\right)\alpha^\mu$$

also becomes small. What does it yield for us? In the numerator of Ξ_0, one might replace $\tilde{\xi}/\tilde{\varepsilon}$ with $\tilde{E}/|\tilde{E}|$. However, it is possible under only one substantial condition.

We are aware that the conversion, for instance, from E to $\tilde{E}$ is accompanied by the replacement of $\mathbf{p}$ with $\mathbf{p} + \hbar\mathbf{k}$. Therefore, the replacement of $\tilde{\xi}/\tilde{\varepsilon}$ with $\tilde{E}/|\tilde{E}|$ is valid strictly for only small values of k, but this region is our interest. To emphasize this point, we introduce the corresponding limit for k. By definition, for $k < mc\delta/\hbar$ we have the region of small energies, and for $k > mc\delta/\hbar$ we have the region of large energies. Quantity $mc^2\delta$ is small relative to the electron rest energy mc^2, such that $\delta \ll 1$. At the same time,

$$mc^2\delta \gg eU \sim \left(\frac{Ze^2}{\hbar c}\right)^2 mc^2;$$

that is, $mc^2\delta$ transcends the energy of a bond of an electron in an atom; recall that Ze is the nuclear charge. Our main problem is the contribution from the region of small energy; then, we briefly discuss the physics of the relativistic contribution from the region $k > mc\delta/\hbar$.

We write the sought difference $Y - Y_0$ in the form

$$Y - Y_0 = \frac{\hbar}{4\pi} \int^{\sim \delta} \alpha_\mu (\Xi - \Xi_0) \alpha^\mu \frac{dk}{k},$$

in which

$$\Xi - \Xi_0 = \left(a + \frac{\tilde{E}}{|\tilde{E}|}\right) \cdot \left(\frac{1}{|\tilde{E}| - aE_\ell + \hbar k} - \frac{1}{\tilde{\varepsilon} - a\varepsilon + \hbar k}\right),$$

and symbol $\sim\delta$ designates the upper bound of k. Note that in the region of small values of k, the denominators in $\Xi - \Xi_0$ become small, but only for $a = 1$. If $a = -1$, then the denominators differ substantially from zero; however, in this case, the numerator might become equal to zero at $\tilde{E} = |\tilde{E}|$. The largest contribution yields the part with $a = 1$. Because our interest is only an approximate nonrelativistic calculation, one might entirely neglect the part with $a = -1$ and replace the sum with respect to a with a simple average. As a result,

$$\Xi - \Xi_0 = \frac{1}{2}\left(1 + \frac{\tilde{E}}{|\tilde{E}|}\right) \cdot \left(\frac{1}{|\tilde{E}| - E_\ell + \hbar k} - \frac{1}{\tilde{\varepsilon} - \varepsilon + \hbar k}\right).$$

For further consideration, we introduce the quantity

$$X = \frac{1}{|\tilde{E}| - E_\ell + \hbar k} - \frac{1}{\tilde{\varepsilon} - \varepsilon + \hbar k}$$

and represent it in two identical ways, namely

$$X(\text{I}) = \frac{1}{\hbar k} - \frac{|\tilde{E}| - E_\ell}{(|\tilde{E}| - E_\ell + \hbar k)\hbar k} - \left(\frac{1}{\hbar k} - \frac{\tilde{\varepsilon} - \varepsilon}{(\tilde{\varepsilon} - \varepsilon + \hbar k)\hbar k}\right)$$

$$= -\frac{|\tilde{E}| - E_\ell}{(|\tilde{E}| - E_\ell + \hbar k)\hbar k} + \frac{\tilde{\varepsilon} - \varepsilon}{(\tilde{\varepsilon} - \varepsilon + \hbar k)\hbar k}$$

and

$$X(\text{II}) = -\frac{|\tilde{E}| - E_\ell}{(\hbar k)^2} + \frac{(|\tilde{E}| - E_\ell)^2}{(\hbar k)^3} - \frac{(|\tilde{E}| - E_\ell)^3}{(|\tilde{E}| - E_\ell + \hbar k)(\hbar k)^3}$$

$$- \left(-\frac{\tilde{\varepsilon} - \varepsilon}{(\hbar k)^2} + \frac{(\tilde{\varepsilon} - \varepsilon)^2}{(\hbar k)^3} - \frac{(\tilde{\varepsilon} - \varepsilon)^3}{(\tilde{\varepsilon} - \varepsilon + \hbar k)(\hbar k)^3}\right),$$

in which, to obtain $X(\text{II})$, we apply twice representation $X(\text{I})$. In $X(\text{II})$, we have greater powers of differences $|\tilde{E}| - E_\ell$ and $\tilde{\varepsilon} - \varepsilon$; moreover, the expressions

$$(\tilde{\varepsilon} - \varepsilon) - (|\tilde{E}| - E_\ell) \text{ and } (\tilde{\varepsilon} - \varepsilon)^2 - (|\tilde{E}| - E_\ell)^2$$

appear.

Let us see what these expressions yield. We have

$$(\tilde{\varepsilon} - \varepsilon) - (|\tilde{E}| - E_\ell) = (\tilde{\varepsilon} - |\tilde{E}|) - (\varepsilon - E_\ell) = \frac{1}{2}(\tilde{\xi}\tilde{F} + \tilde{F}\tilde{\xi}) - \frac{1}{2}(\xi F + F\xi);$$

this quantity is, at most, of order $\hbar k U$. For the difference of squares of the same quantities, we analogously obtain

$$(\tilde{\varepsilon}^2 - |\tilde{E}|^2) + (\varepsilon^2 - E_\ell^2) + 2(|\tilde{E}|E_\ell - \tilde{\varepsilon}\varepsilon)$$
$$= e(\tilde{\xi}\tilde{U} + \tilde{U}\tilde{\xi}) + e(\xi U + U\xi) - [(\tilde{\xi}\tilde{F} + \tilde{F}\tilde{\xi})\varepsilon + \tilde{\varepsilon}(\xi F + F\xi)].$$

With an accuracy up to $\hbar k U$, the expression in square brackets leads to

$$[\xi(F\varepsilon + \varepsilon F) + (F\varepsilon + \varepsilon F)\xi] = 2e(\xi U + U\xi).$$

Consequently, difference $(\tilde{\varepsilon} - \varepsilon)^2 - (|\tilde{E}| - E_\ell)^2$ becomes of order $\hbar k U$. One might generally neglect the contribution from these expressions.

Further, because the momentum of an electron is small, in the expressions for $X(\text{I})$ and $X(\text{II})$, one might then omit the terms containing $\tilde{\varepsilon} - \varepsilon$ and $(\tilde{\varepsilon} - \varepsilon)^3$, respectively. Thus,

$$X(\text{I}) = -\frac{|\tilde{E}| - E_\ell}{(|\tilde{E}| - E_\ell + \hbar k)\hbar k} \text{ and } X(\text{II}) = -\frac{(|\tilde{E}| - E_\ell)^3}{(|\tilde{E}| - E_\ell + \hbar k)(\hbar k)^3}.$$

For which purpose are these two representations of X necessary? It is necessary only when trying to obtain a convenient expression to calculate the Lamb shift. A device to proceed follows. We consider the expression

$$\alpha_\mu (\Xi - \Xi_0) \alpha^\mu$$

and represent it in form

$$\alpha_\mu (\Xi - \Xi_0) \alpha^\mu = \frac{1}{2}\left(1 + \frac{\tilde{E}}{|\tilde{E}|}\right) X(\mathrm{II}) + \frac{1}{2}\alpha_i \left(1 + \frac{\tilde{E}}{|\tilde{E}|}\right) X(\mathrm{I}) \alpha^i.$$

We take into account that α^0 is a simple unit matrix. Note that matrices α^i incarnate the components of a velocity of an electron and, regarding the momentum, their matrix elements become small quantities. We seek to lead the expression

$$\frac{1}{2}\left(1 + \frac{\tilde{E}}{|\tilde{E}|}\right) X(\mathrm{II})$$

to a form that resembles

$$\frac{1}{2}\alpha_i \left(1 + \frac{\tilde{E}}{|\tilde{E}|}\right) X(\mathrm{I}) \alpha^i.$$

This conversion is the device with two expressions for X.
We rewrite the quantity

$$\frac{1}{2}(1 + \tilde{E}/|\tilde{E}|) X(\mathrm{II})$$

in form

$$-(|\tilde{E}| - E_\ell)\left(1 + \frac{\tilde{E}}{|\tilde{E}|}\right) \frac{|\tilde{E}| - E_\ell}{2(\hbar k)^3 (|\tilde{E}| - E_\ell + \hbar k)} (|\tilde{E}| - E_\ell).$$

Then, for $\tilde{E} = -|\tilde{E}|$, the total expression becomes equal to zero. On the right and on the left, one might assume that $|\tilde{E}| - E_\ell = \tilde{E} - E_\ell$. In turn,

$$(\tilde{E} - E_\ell)|\ell\rangle = (\boldsymbol{\alpha} \cdot (\mathbf{p} + \hbar \mathbf{k}) + \beta - eU - E_\ell)|\ell\rangle = \hbar(\boldsymbol{\alpha} \cdot \mathbf{k})|\ell\rangle.$$

What do we see? The diagonal matrix element from

$$\frac{1}{2}\left(1 + \frac{\tilde{E}}{|\tilde{E}|}\right) X(\mathrm{II})$$

equals the diagonal matrix element from

$$-\hbar(\alpha \cdot \mathbf{k})\left(1 + \frac{\tilde{E}}{|\tilde{E}|}\right)\frac{|\tilde{E}| - E_\ell}{2(\hbar k)^3(|\tilde{E}| - E_\ell + \hbar k)}\hbar(\alpha \cdot \mathbf{k}).$$

However, to estimate the Lamb shift, we require only the diagonal elements. Therefore, we focus our attention on the latter expression and calculate the average value with respect to the directions of vector $l = \mathbf{k}/k$. In spherical coordinates, we encounter the trivial integrals

$$\frac{1}{4\pi}\int_0^{2\pi}\int_0^{\pi} \ell^i \ell^j \cdot \sin\theta \, d\theta \, d\phi = \frac{1}{3}\delta_{ij}.$$

As a result,

$$\langle \ell | \frac{1}{2}(1 + \tilde{E}/|\tilde{E}|)X(\text{II})|\ell\rangle = -\langle \ell | \frac{1}{6}\alpha_i(1 + \tilde{E}/|\tilde{E}|)X(\text{I})\alpha^i|\ell\rangle,$$

in which $\alpha_i = -\alpha^i$.

The device is completed; we can return to $\alpha_\mu(\Xi - \Xi_0)\alpha^\mu$. We have

$$\langle \ell | \alpha_\mu(\Xi - \Xi_0)\alpha^\mu|\ell\rangle = \langle \ell | \frac{1}{3}\alpha_i(1 + \tilde{E}/|\tilde{E}|)X(\text{I})\alpha^i|\ell\rangle.$$

For the sought difference $Y - Y_0$, we consequently obtain

$$\langle \ell|(Y - Y_0)|\ell\rangle = -\frac{1}{12\pi}\langle \ell|\int^{\sim\delta} \alpha_i \frac{(1 + E/|E|)(|E| - E_\ell)}{|E| - E_\ell + \hbar k}\alpha^i \frac{d\mathbf{k}}{k^2}|\ell\rangle,$$

in which we replace $\tilde{E}$ with E. This replacement is valid for only small values of $\hbar k$ and for obtaining the first nonzero approximation, whereas in the general case $\tilde{E}$ depends on k. We proceed to calculate the integral in spherical coordinates when $d\mathbf{k} = 4\pi k^2 dk$. Such a calculation yields

$$\langle \ell|(Y - Y_0)|\ell\rangle = -\frac{1}{3\hbar}\langle \ell|\alpha_i\left(1 + \frac{E}{|E|}\right)(|E| - E_\ell)\ln\left(\frac{m\delta + |E| - E_\ell}{||E| - E_\ell|}\right)\alpha^i|\ell\rangle.$$

In the logarithm, we neglect difference $|E| - E_\ell$ relative to $m\delta$; then,

$$\langle \ell|(Y - Y_0)|\ell\rangle = \frac{2}{3\hbar}\sum_{n,i}\langle \ell|\alpha^i|n\rangle(E_n - E_\ell)\ln\left(\frac{m\delta}{|E_n - E_\ell|}\right)\langle n|\alpha^i|\ell\rangle,$$

in which quantity $1 + E/|E|$ has been converted into 2, and the summation is performed with respect to positive values E_n; $\alpha_i = -\alpha^i$.

We finally restore c. For this purpose, recall that on neglecting eU, quantity $c\alpha$ is velocity $\mathbf{p}/m$ of an electron; hence, approximately

$$\langle\ell|\alpha^i|n\rangle \to \frac{\langle\ell|p^i|n\rangle}{cm}.$$

In the logarithm, we replace m with mc, because for k the bound of integration is equal to $(mc/\hbar)\delta$. The energy of Dirac's electron also lacks c, such that

$$E_n - E_\ell \to \frac{E_n - E_\ell}{c}.$$

For the shift, multiplying $\langle\ell|(Y - Y_0)|\ell\rangle$ by e^2/π, we thus eventually obtain

$$\Delta E_\delta = \frac{2e^2}{3\pi\hbar m^2 c^3} \sum_n |\mathbf{p}_{n\ell}|^2 (E_n - E_\ell) \ln\left(\frac{mc^2\delta}{|E_n - E_\ell|}\right).$$

What do we discover? This expression is the Bethe formula, but with a small variation of form. However, $mc^2\delta$ figures in the logarithm, not just mc^2. Because quantity $mc^2\delta$ is supposed to be sufficiently large, shift ΔE_δ differs weakly from the result that is obtainable through the Bethe formula. Parameter δ is somewhat arbitrary; as we already know, it defines the bound between the low frequency and high frequency parts of the Lamb shift. Our formula fails to take into account the high energy contribution from values $mc^2\delta$ and greater. A regularization truncates the part with too great energy, for instance, at value mc^2; as a result, the Lamb shift becomes expressible as

$$\frac{\hbar}{4\pi}\int_0^{mc\delta/\hbar} \alpha_\mu(\Xi - \Xi_0)\alpha^\mu \frac{dk}{k} + \frac{\hbar}{4\pi}\int_{mc\delta/\hbar}^{mc/\hbar} \alpha_\mu(\Xi - \Xi_0)\alpha^\mu \frac{dk}{k},$$

in which the bounds of integration belong to modulus $|\mathbf{k}|$ of that wave vector. We have calculated the first integral; the calculation of the second one generally requires much more effort. This calculation is simple but becomes much too bulky. One must again approximate expression $\Xi - \Xi_0$ to make it suitable for integrating in the high frequency part. These computations bear resemblance to those that we performed to obtain the anomalous magnetic moment of an electron.

It is intuitively clear that, as a result, logarithmic term $\ln(1/\delta)$ appears with the same coefficient as in ΔE_δ before the logarithm. After joining of two integrals, δ consequently vanishes from the eventual solution. This situation is correct. According to Dirac's theory, the nonrelativistic calculation with expression ΔE_δ yields the principal contribution to the Lamb shift. For the sake of interest, let us demonstrate the result $\Delta E'_\delta$ for the electromagnetic shift that is caused by the contribution of the high energy part of the integral in Y:

$$\Delta E'_\delta = \frac{2e^2}{3\pi\hbar m^2 c^3} \sum_n |\mathbf{p}_{n\ell}|^2 (E_n - E_\ell)\left(\ln\left(\frac{1}{2\delta}\right) + \frac{5}{6}\right) - \frac{e^3}{4\pi m^2 c^3}[\boldsymbol{\sigma}\cdot(\operatorname{grad} U \times \mathbf{p})]_{\ell\ell};$$

Dirac calculated this integral. Among the many authors who obtained this formula, French and Weisskopf found the total shift.

Vacuum polarization

The Dirac sea of electrons with negative energy leads not only to an understanding of positrons but also to a peculiar interaction that arises between electric charges. One might suggest that a real electron, which possesses a positive energy, in repulsing from itself the electrons with negative energy polarizes a vacuum around itself. The measured charge and the charge, which is supposed to be true, therefore fail to coincide. Through a polarization of the electron−positron vacuum, a true charge of an electron becomes slightly greater than a measured charge. This effect is especially clear in the spectrum of hydrogen for S-levels that are shifted down slightly more than other energy levels. As a result, the electromagnetic Lamb shift is slightly compensated.

According to the developed Dirac theory, the polarization effects are described through quantity Z with matrix elements

$$Z_{n\ell} = \frac{\hbar}{\pi} \int a J_k^\mu \frac{\langle n|\alpha_\mu e^{-i\mathbf{k}\cdot\mathbf{r}}|\ell\rangle}{E_n - E_\ell + a\hbar k} \cdot \frac{d\mathbf{k}}{k},$$

in which

$$J_k^\mu = \frac{1}{2} \sum_{n'} \langle n'|\alpha^\mu e^{i\mathbf{k}\cdot\mathbf{r}} \eta(-E)|n'\rangle.$$

The validity of this assertion is emphasized at least by the Heaviside function

$$\eta(-E) = \frac{1}{2}\left(1 - \frac{E}{|E|}\right),$$

which becomes equal to zero at $E = |E|$. Quantity Z takes into account the influence of the region of negative values of energy. We already know that Z yields no contribution to the anomalous magnetic moment of an electron; however, it must contribute to the Lamb shift. This contribution is calculable. We proceed to estimate the influence of a vacuum polarization on a shift of atomic levels.

Let us work in the momentum representation. Then,

$$J_k^\mu = \frac{1}{4} \iint \langle \mathbf{p}'|e^{i\mathbf{k}\cdot\mathbf{r}}|\mathbf{p}''\rangle \langle \mathbf{p}''|\mathrm{Sp}\left(\alpha^\mu\left(1 - \frac{E}{|E|}\right)\right)|\mathbf{p}'\rangle d\mathbf{p}' \, d\mathbf{p}'',$$

in which $\langle \mathbf{p}'|e^{i\mathbf{k}\cdot\mathbf{r}}|\mathbf{p}''\rangle = \delta(\mathbf{p}' - \mathbf{p}'' - \hbar\mathbf{k})$. Moreover, because one must calculate the sum of diagonal elements $\langle n'|...|n'\rangle$ with respect to all possible states, we add the spur with respect to the spin variables. From the preceding section we know that, in the absence of a static magnetic field,

$$1 - \frac{E}{|E|} = 1 - \frac{\xi}{\varepsilon} + \frac{1}{\varepsilon}K\frac{1}{\varepsilon},$$

in which

$$K = \frac{1}{2}(\varepsilon F \varepsilon - \xi F \xi).$$

Consequently,

$$J_k^\mu = \frac{1}{4}\iint \delta(\mathbf{p}' - \mathbf{p}'' - \hbar\mathbf{k})\langle\mathbf{p}''|\mathrm{Sp}\left(\alpha^\mu\left(1 - \frac{\xi}{\varepsilon} + \frac{1}{\varepsilon}K\frac{1}{\varepsilon}\right)\right)|\mathbf{p}'\rangle d\mathbf{p}' \, d\mathbf{p}''.$$

Element $\langle\mathbf{p}''|\mathrm{Sp}(\alpha^\mu(1-\xi/\varepsilon))|\mathbf{p}'' + \hbar\mathbf{k}\rangle$ differs from zero only at $\mathbf{k} = 0$, but this case, as we are already convinced, should be excluded from consideration. Therefore, there remains

$$J_k^\mu = \frac{1}{4}\iint \delta(\mathbf{p}' - \mathbf{p}'' - \hbar\mathbf{k})\langle\mathbf{p}''|\frac{1}{\varepsilon}\mathrm{Sp}(\alpha^\mu K)\frac{1}{\varepsilon}|\mathbf{p}'\rangle d\mathbf{p}' \, d\mathbf{p}''.$$

Let us calculate spurs

$$\mathrm{Sp}(\alpha^0 K) = \frac{1}{2}\mathrm{Sp}(\alpha^0(\varepsilon F \varepsilon - \xi F \xi)) \text{ and } \mathrm{Sp}(\alpha^i K) = \frac{1}{2}\mathrm{Sp}(\alpha^i(\varepsilon F \varepsilon - \xi F \xi)).$$

Through these equalities

$$\mathrm{Sp}(\alpha^i) = 0, \quad \mathrm{Sp}(\beta) = 0, \quad \mathrm{Sp}(\alpha^i\beta) = 0, \text{ and } \mathrm{Sp}(\alpha^i\alpha^j\alpha^k) = 0,$$

the second spur equals zero. To calculate the first spur, we note that

$$2K = \varepsilon F \varepsilon - [(\boldsymbol{\alpha} \cdot \mathbf{p})F(\boldsymbol{\alpha} \cdot \mathbf{p}) + (\boldsymbol{\alpha} \cdot \mathbf{p})F\beta + \beta F(\boldsymbol{\alpha} \cdot \mathbf{p}) + m^2 F],$$

and also $\mathrm{Sp}(\alpha^0) = 4$ and

$$\mathrm{Sp}[(\boldsymbol{\alpha} \cdot \mathbf{p})\alpha^0 F(\boldsymbol{\alpha} \cdot \mathbf{p})] = \mathbf{p} \cdot \mathrm{Sp}(\alpha^0 F) \cdot \mathbf{p} = 4\mathbf{p}F\mathbf{p}.$$

Thus, $\mu = 0$ and

$$\mathrm{Sp}(\alpha^0 K) = 2(\varepsilon F \varepsilon - \mathbf{p}F\mathbf{p} - m^2 F).$$

Quantity J_k^μ has only one nonzero component that is

$$J_k^0 = \frac{e}{\hbar^3}\iint \delta(\mathbf{p}'-\mathbf{p}''-\hbar\mathbf{k})U_{(\mathbf{p}'-\mathbf{p}'')/\hbar}\frac{\varepsilon''\varepsilon'-\mathbf{p}''\mathbf{p}'-m^2}{\varepsilon''\varepsilon'(\varepsilon''+\varepsilon')}\,\mathrm{d}\mathbf{p}'\,\mathrm{d}\mathbf{p}'',$$

in which we take into account that

$$\langle\mathbf{p}''|F|\mathbf{p}'\rangle = \frac{2e}{\hbar^3(\varepsilon''+\varepsilon')}U_{(\mathbf{p}'-\mathbf{p}'')/\hbar}.$$

We proceed to integrate with respect to one momentum, for instance, $\mathbf{p}'$. Instead of $\mathbf{p}''$, through the relations

$$\mathbf{p}' = \mathbf{p} + \frac{\hbar\mathbf{k}}{2} \text{ and } \mathbf{p}'' = \mathbf{p} - \frac{\hbar\mathbf{k}}{2},$$

one should introduce a new quantity $\mathbf{p}$. As a result,

$$J_k^0 \equiv \frac{eU_k}{\hbar^3}I_k = \frac{eU_k}{\hbar^3}\int\frac{\varepsilon''\varepsilon'-\mathbf{p}^2+\hbar^2 k^2/4 - m^2}{\varepsilon''\varepsilon'(\varepsilon''+\varepsilon')}\,\mathrm{d}\mathbf{p}.$$

Here, I_k designates the integral that we must calculate.

For the sake of convenience of integration, we choose vector $\mathbf{k}$ to have form $(0,0,k)$ and apply cylindrical coordinates for $\mathbf{p}$, that is,

$$p_x = \rho\cos\phi,\ p_y = \rho\sin\varphi,\ p_z = p_z, \text{ and } \mathrm{d}\mathbf{p} = \rho\,\mathrm{d}\rho\,\mathrm{d}\phi\,\mathrm{d}p_z.$$

Then

$$\varepsilon' = \sqrt{\rho^2 + p_z^2 + \hbar p_z k + \frac{\hbar^2}{4}k^2 + m^2} \text{ and } \varepsilon'' = \sqrt{\rho^2 + p_z^2 - \hbar p_z k + \frac{\hbar^2}{4}k^2 + m^2}.$$

Instead of ρ, we introduce a new variable $f = \varepsilon' + \varepsilon''$, and as

$$\left(\frac{\mathrm{d}f}{\mathrm{d}\rho}\right)_{\substack{p_z=\text{const}\\ \phi=\text{const}}} = \frac{\rho}{\varepsilon'} + \frac{\rho}{\varepsilon''} = \frac{\rho f}{\varepsilon'\varepsilon''},$$

then

$$\rho\,\mathrm{d}\rho\,\mathrm{d}\phi\,\mathrm{d}p_z = \varepsilon'\varepsilon''\frac{\mathrm{d}f}{f}\,\mathrm{d}\phi\,\mathrm{d}p_z.$$

Let us transform the integrand:

$$\frac{\varepsilon''\varepsilon' - \mathbf{p}^2 + \hbar^2 k^2/4 - m^2}{\varepsilon''\varepsilon'(\varepsilon'' + \varepsilon')} = \frac{1}{f\varepsilon'\varepsilon''}\left(\varepsilon'\varepsilon'' - \frac{1}{2}(\varepsilon'^2 + \varepsilon''^2 - \hbar^2 k^2)\right)$$

$$= \frac{1}{2f\varepsilon'\varepsilon''}(\hbar^2 k^2 - (\varepsilon' - \varepsilon'')^2).$$

Using

$$(\varepsilon' - \varepsilon'')^2 = \left[\frac{\varepsilon'^2 - \varepsilon''^2}{f}\right]^2 = \left(\frac{2\hbar p_z k}{f}\right)^2,$$

we obtain

$$I_k = \pi \hbar^2 k^2 \iint \left(1 - \frac{4p_z^2}{f^2}\right)\frac{df}{f^2}\,dp_z,$$

in which we have already performed the simple integration with respect to angle ϕ. Further, the integration with respect to p_z yields

$$p_z\Big|_{-p_0}^{p_0} = 2p_0 \quad \text{and} \quad \frac{1}{3}p_z^3\Big|_{-p_0}^{p_0} = \frac{2}{3}p_0^3,$$

in which p_0 is the maximum value of p_z. To find value p_0, we put $\rho = 0$ and $p_z = p_0$ in the expression for f^2; namely,

$$f^2 = (\varepsilon' + \varepsilon'')^2 = 2p_0^2 + \frac{\hbar^2}{2}k^2 + 2m^2$$

$$+ 2\sqrt{p_0^2 + \frac{\hbar^2}{4}k^2 + m^2 + \hbar p_0 k}\sqrt{p_0^2 + \frac{\hbar^2}{4}k^2 + m^2 - \hbar p_0 k},$$

therefore

$$\left(p_0^2 + \frac{\hbar^2}{4}k^2 + m^2 - f^2/2\right)^2 = \left(p_0^2 + \frac{\hbar^2}{4}k^2 + m^2\right)^2 - \hbar^2 k^2 p_0^2$$

and

$$p_0 = \frac{f}{2}\sqrt{\frac{f^2 - \hbar^2 k^2 - 4m^2}{f^2 - \hbar^2 k^2}}.$$

As a result,

$$I_k = \pi\hbar^2 k^2 \int \left(\left(\frac{f^2 - \hbar^2 k^2 - 4m^2}{f^2 - \hbar^2 k^2} \right)^{1/2} - \frac{1}{3} \left(\frac{f^2 - \hbar^2 k^2 - 4m^2}{f^2 - \hbar^2 k^2} \right)^{3/2} \right) \frac{df}{f}.$$

The integration over f needs to be performed, so one must preliminarily transform the integral. With the aid of relations

$$x^2 = f^2 - \hbar^2 k^2 \quad \text{and} \quad x\,dx = f\,df,$$

we produce the substitution; obviously,

$$I_k = \frac{2\pi\hbar^2 k^2}{3} \int \frac{\sqrt{x^2 - 4m^2}(x^2 + 2m^2)}{x^2(x^2 + \hbar^2 k^2)} dx.$$

To eliminate the irrational expression, we use

$$y^2 = 1 - \frac{4m^2}{x^2};$$

then,

$$y\,dy = \frac{4m^2}{x^3} dx$$

and

$$I_k = \frac{4\pi m^2 \hbar^2 k^2}{3} \int \frac{(3 - y^2)y^2 dy}{(4m^2 + \hbar^2 k^2 - \hbar^2 k^2 y^2)(1 - y^2)}.$$

Expanding the integrand, we represent it in the form

$$\frac{(4m^2 + \hbar^2 k^2)}{(4m^2 + \hbar^2 k^2 - \hbar^2 k^2 y^2)} \cdot \left(\frac{1}{\hbar^2 k^2} - \frac{1}{2m^2} \right) + \frac{1}{2m^2(1 - y^2)} - \frac{1}{\hbar^2 k^2}.$$

Consequently,

$$I_k = \frac{4\pi m^2}{3}(4m^2 + \hbar^2 k^2)\left(1 - \frac{\hbar^2 k^2}{2m^2}\right) \int \frac{dy}{4m^2 + \hbar^2 k^2 - \hbar^2 k^2 y^2}$$

$$+ \frac{2\pi\hbar^2 k^2}{3} \int \frac{dy}{1 - y^2} - \frac{4\pi m^2}{3} \int dy.$$

Using $\lambda = \hbar^2 k^2/4\, m^2$, we perform the simple integration; as a result,

$$I_k = \frac{2\pi m^2}{3}(1-2\lambda)\sqrt{\frac{1+\lambda}{\lambda}}\ln\left|\frac{\sqrt{1+\lambda}+\sqrt{\lambda y}}{\sqrt{1+\lambda}-\sqrt{\lambda y}}\right| + \frac{4\pi m^2}{3}\left(\lambda\ln\left|\frac{1+y}{1-y}\right| - y\right).$$

Let us elucidate the choice of the bounds of integration. At $\rho = p_z = 0$, quantity f^2 has a minimum value equal to

$$4\,m^2 + \hbar^2 k^2.$$

A regularization sets the maximum value for f^2 that amounts to

$$4\hbar^2\chi^2.$$

For x^2, the lower and upper bounds are consequently equal to

$$4\,m^2 \text{ and } 4\hbar^2\chi^2,$$

respectively. Finally,

$$0 \le y \le 1 - \frac{m^2}{\hbar^2\chi^2}.$$

Substituting the bounds, we see that $I_k = 0$ at $y = 0$. Further,

$$\ln\left|\frac{1+y}{1-y}\right|_{y=1-m^2/\hbar^2\chi^2} = 2\ln\left(\frac{\sqrt{2}\hbar\chi}{m}\right);$$

assuming $y = 1$, we have

$$\ln\left|\frac{\sqrt{1+\lambda}+\sqrt{\lambda}}{\sqrt{1+\lambda}-\sqrt{\lambda}}\right| = \ln\left(\sqrt{\lambda}+\sqrt{1+\lambda}\right)^2.$$

Thus,

$$I_k = \frac{4\pi m^2}{3}\left(2\lambda\ln\left(\frac{\sqrt{2}\hbar\chi}{m}\right) - 1 + (1-2\lambda)\sqrt{\frac{1+\lambda}{\lambda}}\ln\left(\sqrt{\lambda}+\sqrt{1+\lambda}\right)\right).$$

We have generally obtained the required result. Returning to the expression for J_k^0 and then proceeding to $Z_{n\ell}$ to integrate with respect to k still remain. Note that to estimate the contribution to the Lamb shift, we require only the diagonal matrix element

$$Z_{\ell\ell} = \frac{1}{\pi}\int J_k^0 \langle\ell|e^{-i\mathbf{k}\cdot\mathbf{r}}|\ell\rangle \frac{d\mathbf{k}}{k^2}.$$

Taking into account the vacuum polarization, $Z_{\ell\ell}$ represents the variation of energy. Moreover, the region of low frequencies has practical interest; for the hydrogen-type atoms, quantity $\hbar k$ is small relative to m. Consequently, λ becomes a small quantity; one might expand I_k in a series with respect to λ, retaining only the first powers. In this case,

$$(1+\lambda)^{1/2} = 1 + \frac{\lambda}{2} - \frac{\lambda^2}{8}$$

and

$$\ln\left(\sqrt{\lambda} + \sqrt{\lambda+1}\right) = \ln\left(1 + \left(\sqrt{\lambda} + \frac{\lambda}{2} - \frac{\lambda^2}{8}\right)\right) = \sqrt{\lambda} - \frac{1}{6}\lambda^{3/2} + \frac{3}{40}\lambda^{5/2}.$$

Note that for the logarithm, one must maintain terms up to the fifth order with respect to $\sqrt{\lambda}$. With an accuracy up to λ^2, we have

$$I_k = \frac{4\pi m^2}{3}\left(2\lambda \ln\left(\frac{\sqrt{2}\hbar\chi}{m}\right) - \frac{5}{3}\lambda - \frac{4}{5}\lambda^2\right);$$

hence,

$$J_k^0 = \frac{q'e}{\hbar}U_k k^2 - \frac{\pi e\hbar}{15\,m^2}U_k k^4,$$

in which

$$q' = \frac{2\pi}{3}\left(\ln\left(\frac{\sqrt{2}\hbar\chi}{m}\right) - \frac{5}{6}\right).$$

Let us calculate $Z_{\ell\ell}$. Assuming that $\langle \ell | Z | \ell \rangle$ is the matrix element of quantity Z, we write

$$Z = \frac{e}{\pi}\int\left(\frac{q'}{\hbar} - \frac{\pi\hbar}{15\,m^2}k^2\right)U_k e^{-i\mathbf{k}\cdot\mathbf{r}}d\mathbf{k}.$$

However,

$$\int U_k e^{-i\mathbf{k}\cdot\mathbf{r}}d\mathbf{k} = U(\mathbf{r}) \quad \text{and} \quad \int k^2 U_k e^{-i\mathbf{k}\cdot\mathbf{r}}d\mathbf{k} = -\nabla^2 U(\mathbf{r});$$

hence,

$$Z = \frac{eq'}{\pi\hbar}U + \frac{e\hbar}{15\,m^2}\nabla^2 U.$$

The sought variation of energy ΔE_{pol}, which is equal to $e^2 Z_{\ell\ell}/\pi$, acquires the form

$$\Delta E_{\text{pol}} = \frac{e^3}{\pi^2 \hbar} q' U_{\ell\ell} + \frac{e^3 \hbar}{15\pi m^2} (\nabla^2 U)_{\ell\ell}.$$

Restoring c, we obtain

$$\Delta E_{\text{pol}} = \frac{q'}{\pi^2}\left(\frac{e^2}{\hbar c}\right) e U_{\ell\ell} + \frac{\hbar^2}{15\pi (mc)^2}\left(\frac{e^2}{\hbar c}\right) e (\nabla^2 U)_{\ell\ell},$$

in which

$$q' = \frac{2\pi}{3}\left(\ln\left(\frac{\sqrt{2}\hbar\chi}{mc}\right) - \frac{5}{6}\right).$$

Renormalization of charge

The obtained quantity ΔE_{pol} defines the contribution from the polarization of the vacuum to an electromagnetic shift of atomic levels. However, one remark is pertinent. Quantity q', which appears in ΔE_{pol}, might be great, especially for large values of χ. We already know that a similar situation arises when considering a free electron; to solve that problem, we applied the procedure of a renormalization of mass. It turns out that one might also eliminate q'. For this purpose, we apply the procedure of renormalization of charge. In an initial Hamiltonian, the charge is some effective parameter to which we ascribe value $e + \delta e$. We understand e as an observable charge, whereas $\delta e \sim e(e^2/\hbar c)$ enters into the additional Hamiltonian

$$\delta H = -\delta e \int \psi^+ U \psi \, d\mathbf{r} = -\delta e \sum_{nn'} \varphi_n^+ \langle n|U|n'\rangle \varphi_{n'},$$

which is considered as a perturbation of the second order of smallness. The presence of δH leads to $\varphi_{\ell(2)}^+$ for the additional variation that is determined by the equation,

$$i\hbar \frac{\partial}{\partial t}\varphi_{\ell(2)}^+ = [\delta H', \varphi_\ell^+].$$

Proceeding to the interaction picture from δH to $\delta H'$, we replace φ_n^+ and $\varphi_{n'}$ in δH with $e^{iE_n t/\hbar}\varphi_n^+$ and $e^{-iE_{n'}t/\hbar}\varphi_{n'}$, respectively. As a result,

$$i\hbar \frac{\partial}{\partial t}\varphi_{\ell(2)}^+ = -\delta e \sum_n \varphi_n^+ \langle n|U|\ell\rangle e^{i(E_n - E_\ell)t/\hbar}.$$

Here, the term with $n = \ell$ represents the additional variation of energy of the second order. Adding it to ΔE_{pol}, we equate the expression with $U_{\ell\ell}$ to zero. Hence,

$$\frac{\delta e}{e} = \frac{q'}{\pi^2}\left(\frac{e^2}{\hbar c}\right).$$

This result is the renormalization of charge. The numerical estimations, which were performed for the case of renormalization of mass, are approximately maintained here.

The real contribution to the Lamb shift becomes equal to

$$\frac{\hbar e^3}{15\pi m^2 c^3}(\nabla^2 U)_{\ell\ell}.$$

To obtain the total shift, one must add it to the result that was found for an electromagnetic shift of levels in the preceding section.

Dirac's ideas and quantum field theory

Discussing the fields, we imply real physical objects, for instance, the electrons and photons. At present, field formalism extends far beyond the relativistic quantum mechanics: it is applicable in the physics of solids, in the theory of an atomic nucleus, and in the theory of a plasma, and also in many other branches of physics to describe the principal characteristics of quantum systems. It is well known that one might ascribe a field of phonons to the vibrations of a crystalline lattice; to describe ferromagnetic phenomena, one should apply a concept of spin waves that represent a field of magnons. One might continue this series, but it is already clear that the method of quantum fields is firmly entrenched in a prime place in all quantum physics. Even for some problems of the simplest atoms such as hydrogen and helium, it becomes simpler to apply a many-particle description than to apply a clear traditional representation of a system in the form of several interacting particles. Through the fields we comprehend the particles.

Our aim was to generate a first acquaintance with the field point of view that allows one to enter the general range of questions of quantum electrodynamics. In the Heisenberg picture, we have obtained information about an electron when considering the temporal variation of its creation operator. One might perform something similar for a creation operator of a photon and, further, for example, for an operator of creation of a muon under the condition that it becomes possible to introduce a muon field into the theory. However, we fail to derive special dividends. For instance, the self-energy of a photon becomes infinite. Yes, we might apply the procedure of a regularization to ensure that this energy becomes large but finite; however, we cannot eliminate it entirely. If a Hamiltonian comprises a mass parameter, then we can manipulate it, suggesting that an electromagnetic part of a mass, a meson part, and others exist. For a photon, this scenario is questionable.

Beginning with the Bethe formula, many works aiming at seeking practical rules of a game with expressions that generally diverge appeared, but not a general comprehensive theory. We concentrated our attention on the theory that was developed by Dirac, which is, in our opinion, suitable for a first acquaintance with the principles of quantum electrodynamics. Despite the fact that there was a common preference to work with Schrödinger's formalism, Dirac believed that it was much more logical to use Heisenberg's equations. In this case, the theory retains the natural harmony and a reasonable sequence of conclusions.

This theory certainly does not lack shortcomings. First, we failed to cope with ultraviolet divergences; as a result of the procedure of regularization of divergent integrals, the theory lost relativistic invariance. Second, in the calculations of observables, we neglected some quantities that were supposed to be small; however, this condition is not entirely obvious and many approximations are questionable. Third, a gauge-invariant Hamiltonian was formally separated into two parts that are related to the free fields and their interaction, respectively; the second part was considered a perturbation. We have placed both parts on unequal footing and partly lost the sense of gauge invariance.

In other respects, as we have seen, the theory is satisfactory. Solving Heisenberg's equations, Dirac obtained general quantities Y and Z that describe the interaction of fields according to the perturbation theory. A regularization organically enters a computation, excluding the region of high energies from consideration. Various approximations are necessary to achieve an agreement with the experiment. How should we understand such an approach? Likely, the theory cannot be perfect, so subsidiary rules of a game become necessary. We ultimately arrive at the results in a form of expansions with respect to a coupling constant. This form is convenient for a comparison with the experiment, but at the same time it forces us to work in a framework of the method of the perturbation theory. One might think that, in the course of the calculations, we instinctively neglect some quantities only to convert our solution into a form resembling a reasonable expansion with respect to a coupling constant. Our actions are justifiable because this constant is small. If it becomes possible to represent a solution in the form of some function of coupling constant without applying perturbation theory, then a question regarding a gauge invariance of separate parts of a Hamiltonian becomes removed. One might conclude that quantum electrodynamics is based on not only the equations and the methods to solve them but also the definite rules of the game to operate with the field variables.

Let us note that such a situation also arises in any other formulation of quantum field theory. By that we mean the theory of a scattering matrix that is widespread at present. In contrast to Dirac's ideas, it is based on Schrödinger's equation

$$i\hbar \frac{\partial}{\partial t} |\psi\rangle = (H_0 + H_{\text{int}})|\psi\rangle$$

for states $|\psi(t)\rangle$. This equation is also considered in the interaction picture, that is,

$$|\psi\rangle \to e^{-iH_0 t/\hbar}|\psi\rangle, \quad H'_{\text{int}} = e^{iH_0 t/\hbar} H_{\text{int}} e^{-iH_0 t/\hbar}$$

and

$$i\hbar \frac{\partial}{\partial t}|\psi\rangle = H'_{int}(t)|\psi\rangle.$$

A device is as follows. Increment

$$|\psi(t+\delta t)\rangle - |\psi(t)\rangle$$

is represented in the form $(-i\delta t H'_{int}/\hbar)|\psi(t)\rangle$, hence

$$|\psi(t+\delta t)\rangle = e^{-i\delta t H'_{int}/\hbar}|\psi(t)\rangle.$$

One might express $|\psi(t)\rangle$ at an arbitrary moment of time, knowing the value of $|\psi(t)\rangle$ at moment t_0. We have

$$|\psi(t)\rangle = \left(\lim_{\delta t \to 0} \prod_s e^{-i\delta t_s H'_{int}(t_s)/\hbar}\right)|\psi(t_0)\rangle.$$

Through the noncommutativity of H'_{int}, one must retain an initial order for all factors. For this purpose, it is convenient to introduce the chronological Dyson operator T. As a result, setting t_0 equal to $-\infty$ and t to $+\infty$, we obtain

$$|\psi(+\infty)\rangle = S|\psi(-\infty)\rangle,$$

in which

$$S = T e^{-(i/\hbar) \int_{-\infty}^{+\infty} H'_{int}(t)dt};$$

S is a scattering matrix. To apply a perturbation theory, one should represent it in the form of an expansion in powers of an interaction energy:

$$S = \sum_{s=0}^{\infty} \left(-\frac{i}{\hbar}\right)^s \frac{1}{s!} \int_{-\infty}^{+\infty} dt_1 \int_{-\infty}^{+\infty} dt_2 \ldots \int_{-\infty}^{+\infty} dt_s T\left[H'_{int}(t_1)H'_{int}(t_2)\ldots H'_{int}(t_s)\right].$$

Operator T organizes quantities H'_{int} from right to left in order of increasing time t. Note that

$$H'_{int} = \int \ldots d\mathbf{r};$$

hence,

$$S = T e^{-(i/\hbar) \int \ldots (d\mathbf{r}\, dt)},$$

and Lorentz invariance of S becomes obvious.

One sees that quantity S transforms a state of a system before a collision at moment $t = -\infty$ into a state far in the future when all particles are again free but have already interacted. The matrix elements of S correspond to the real transitions that conform to the laws of conservation of energy and momentum. We thus are concerned with only observables. The latter is a plus factor but all aforementioned problems remain and, in addition, new problems arise. The concept of a scattering matrix, first enunciated by Wheeler and independently Heisenberg, knowing a state of a system before an interaction, yields the possibility to determine the probabilities of final states. For instance, let an initial state be |electron + positron⟩ and let a final state be |two photons⟩. Having calculated the matrix element from S between these states, one might realize whether such a process is possible in principle. This scheme is convenient from a practical point of view.

Dirac disagreed with such a formulation. The point is that the Heisenberg and Schrödinger methods of description, which are equivalent in quantum mechanics, become distinguishable in quantum electrodynamics. Let us consider the transfer from one picture to another. State vector |Sc⟩ in the Schrödinger picture is related to Heisenberg's vector |He⟩ through relation

$$|\text{Sc}\rangle = e^{-iHt/\hbar} |\text{He}\rangle.$$

Having obtained a solution, for instance, for vector |He⟩, one might think that |Sc⟩ also exists. However, this supposition is not entirely true. One might assert that it is true only under the condition that quantity $e^{-iHt/\hbar}$ is defined correctly, whereas that cannot be accomplished in quantum electrodynamics. We know that the Hamiltonian of free fields is characterized by eigenvalues with infinite constants, and the energy of interaction demands regularization. The interaction is so great, especially in the high frequency region, that there is no assurance of a possibility to proceed from |He⟩ to |Sc⟩ through multiplication by $e^{-iHt/\hbar}$.

To approach this question from a formal point of view, one might assume that Schrödinger's vector exists but has no coordinates. According to Dirac, such a vector belongs to the space with a dimension that is too large. The interaction is so intense that vector |Sc⟩ leaves a Hilbert space for an infinitesimal time. Working with the Heisenberg equations, we encounter something similar. We fail to assert that the dynamical variables act on the vectors belonging exclusively to a Hilbert space. In this picture, we are nevertheless able to retain the consistency of the theory in the language of q-numbers and have brought it closer to the analogue of classical theory.

The physics of q-numbers and the collective image of a scattering matrix are two points of view of one and the same. Which is more correct? Which is more natural? The answer, most likely, is the theory in the language of dynamical variables, because the observables belong to the people whereas the quantum theory belongs to the world of q-numbers. The principal problem is how to extract the information from this abstract world in such a way that does not spoil the theory and that satisfies the experiment.

Bibliography

1. Blokhintsev DI. *Principles of quantum mechanics*. Moscow: Nauka; 1976.
2. Dirac PAM. *The principles of quantum mechanics*. Oxford: Clarendon Press; 1958.
3. Kirzhnits DA. *Formulation of quantum theory based on differentiation with respect to coupling parameter. Problems of theoretical physics: a volume dedicated to the memory of Igor E. Tamm*. Moscow: Nauka; 1972.
4. Green HS. *Matrix mechanics*. Groningen: Noordhoff; 1965.
5. Kazakov KV. *Quantum theory of anharmonic effects in molecules*. Milton Keynes: Elsevier; 2012.
6. Kazakov KV. Electro-optics of molecules. *Opt Spectrosc* 2004;**97**:725–34.
7. Kazakov KV. Electro-optics of molecules. II. *Opt Spectrosc* 2008;**104**:477–90.
8. Kazakov KV. Formalism of quantum number polynomials. *Russ Phys J* 2005;**48**:954–65.
9. Smith MAH, Rinsland CP, Fridovich B, Rao KN. Intensities and collision broadening parameters from infrared spectra. In: Rao KN, editor. *Molecular spectroscopy: modern research*. New York, NY: Academic Press; 1985. p. 111–248.
10. Ogilvie JF. *The vibrational and rotational spectrometry of diatomic molecules*. London: Academic Press; 1998.
11. Wei H. Four-parameter exactly solvable potential for diatomic molecules. *Phys Rev A* 1990;**42**:2524–9.
12. Bethe HA. *Intermediate quantum mechanics*. New York, NY: W.A. Benjamin Inc.; 1964.
13. Dirac PAM. *Lectures on quantum field theory*. New York, NY: Belfer Graduate School of Science, Yeshiva University; 1966.
14. Fermi E. Quantum theory of radiation. *Rev Mod Phys* 1932;**4**:87–132.
15. Bethe HA. The electromagnetic shift of energy levels. *Phys Rev* 1947;**72**:339–41.

哈尔滨工业大学出版社刘培杰数学工作室
已出版(即将出版)图书目录

书　　名	出版时间	定　价	编号
新编中学数学解题方法全书(高中版)上卷	2007—09	38.00	7
新编中学数学解题方法全书(高中版)中卷	2007—09	48.00	8
新编中学数学解题方法全书(高中版)下卷(一)	2007—09	42.00	17
新编中学数学解题方法全书(高中版)下卷(二)	2007—09	38.00	18
新编中学数学解题方法全书(高中版)下卷(三)	2010—06	58.00	73
新编中学数学解题方法全书(初中版)上卷	2008—01	28.00	29
新编中学数学解题方法全书(初中版)中卷	2010—07	38.00	75
新编中学数学解题方法全书(高考复习卷)	2010—01	48.00	67
新编中学数学解题方法全书(高考真题卷)	2010—01	38.00	62
新编中学数学解题方法全书(高考精华卷)	2011—03	68.00	118
新编平面解析几何解题方法全书(专题讲座卷)	2010—01	18.00	61
新编中学数学解题方法全书(自主招生卷)	2013—08	88.00	261
数学眼光透视	2008—01	38.00	24
数学思想领悟	2008—01	38.00	25
数学应用展观	2008—01	38.00	26
数学建模导引	2008—01	28.00	23
数学方法溯源	2008—01	38.00	27
数学史话览胜	2008—01	28.00	28
数学思维技术	2013—09	38.00	260
从毕达哥拉斯到怀尔斯	2007—10	48.00	9
从迪利克雷到维斯卡尔迪	2008—01	48.00	21
从哥德巴赫到陈景润	2008—05	98.00	35
从庞加莱到佩雷尔曼	2011—08	138.00	136
数学奥林匹克与数学文化(第一辑)	2006—05	48.00	4
数学奥林匹克与数学文化(第二辑)(竞赛卷)	2008—01	48.00	19
数学奥林匹克与数学文化(第二辑)(文化卷)	2008—07	58.00	36′
数学奥林匹克与数学文化(第三辑)(竞赛卷)	2010—01	48.00	59
数学奥林匹克与数学文化(第四辑)(竞赛卷)	2011—08	58.00	87
数学奥林匹克与数学文化(第五辑)	2015—06	98.00	370

哈尔滨工业大学出版社刘培杰数学工作室
已出版(即将出版)图书目录

书　名	出版时间	定　价	编号
世界著名平面几何经典著作钩沉——几何作图专题卷(上)	2009-06	48.00	49
世界著名平面几何经典著作钩沉——几何作图专题卷(下)	2011-01	88.00	80
世界著名平面几何经典著作钩沉(民国平面几何老课本)	2011-03	38.00	113
世界著名平面几何经典著作钩沉(建国初期平面三角老课本)	2015-08	38.00	507
世界著名解析几何经典著作钩沉——平面解析几何卷	2014-01	38.00	264
世界著名数论经典著作钩沉(算术卷)	2012-01	28.00	125
世界著名数学经典著作钩沉——立体几何卷	2011-02	28.00	88
世界著名三角学经典著作钩沉(平面三角卷Ⅰ)	2010-06	28.00	69
世界著名三角学经典著作钩沉(平面三角卷Ⅱ)	2011-01	38.00	78
世界著名初等数论经典著作钩沉(理论和实用算术卷)	2011-07	38.00	126
发展空间想象力	2010-01	38.00	57
走向国际数学奥林匹克的平面几何试题诠释(上、下)(第1版)	2007-01	68.00	11,12
走向国际数学奥林匹克的平面几何试题诠释(上、下)(第2版)	2010-02	98.00	63,64
平面几何证明方法全书	2007-08	35.00	1
平面几何证明方法全书习题解答(第1版)	2005-10	18.00	2
平面几何证明方法全书习题解答(第2版)	2006-12	18.00	10
平面几何天天练上卷·基础篇(直线型)	2013-01	58.00	208
平面几何天天练中卷·基础篇(涉及圆)	2013-01	28.00	234
平面几何天天练下卷·提高篇	2013-01	58.00	237
平面几何专题研究	2013-07	98.00	258
最新世界各国数学奥林匹克中的平面几何试题	2007-09	38.00	14
数学竞赛平面几何典型题及新颖解	2010-07	48.00	74
初等数学复习及研究(平面几何)	2008-09	58.00	38
初等数学复习及研究(立体几何)	2010-06	38.00	71
初等数学复习及研究(平面几何)习题解答	2009-01	48.00	42
几何学教程(平面几何卷)	2011-03	68.00	90
几何学教程(立体几何卷)	2011-07	68.00	130
几何变换与几何证题	2010-06	88.00	70
计算方法与几何证题	2011-06	28.00	129
立体几何技巧与方法	2014-04	88.00	293
几何瑰宝——平面几何500名题暨1000条定理(上、下)	2010-07	138.00	76,77
三角形的解法与应用	2012-07	18.00	183
近代的三角形几何学	2012-07	48.00	184
一般折线几何学	2015-08	48.00	203
三角形的五心	2009-06	28.00	51
三角形的六心及其应用	2015-10	68.00	542
三角形趣谈	2012-08	28.00	212
解三角形	2014-01	28.00	265
三角学专门教程	2014-09	28.00	387

哈尔滨工业大学出版社刘培杰数学工作室
已出版(即将出版)图书目录

书　　名	出版时间	定　价	编号
距离几何分析导引	2015—02	68.00	446
圆锥曲线习题集(上册)	2013—06	68.00	255
圆锥曲线习题集(中册)	2015—01	78.00	434
圆锥曲线习题集(下册)	即将出版		
近代欧氏几何学	2012—03	48.00	162
罗巴切夫斯基几何学及几何基础概要	2012—07	28.00	188
罗巴切夫斯基几何学初步	2015—06	28.00	474
用三角、解析几何、复数、向量计算解数学竞赛几何题	2015—03	48.00	455
美国中学几何教程	2015—04	88.00	458
三线坐标与三角形特征点	2015—04	98.00	460
平面解析几何方法与研究(第1卷)	2015—05	18.00	471
平面解析几何方法与研究(第2卷)	2015—06	18.00	472
平面解析几何方法与研究(第3卷)	2015—07	18.00	473
解析几何研究	2015—01	38.00	425
解析几何学教程.上	2016—01	38.00	574
解析几何学教程.下	2016—01	38.00	575
几何学基础	2016—01	58.00	581
初等几何研究	2015—02	58.00	444
俄罗斯平面几何问题集	2009—08	88.00	55
俄罗斯立体几何问题集	2014—03	58.00	283
俄罗斯几何大师——沙雷金论数学及其他	2014—01	48.00	271
来自俄罗斯的5000道几何习题及解答	2011—03	58.00	89
俄罗斯初等数学问题集	2012—05	38.00	177
俄罗斯函数问题集	2011—03	38.00	103
俄罗斯组合分析问题集	2011—01	48.00	79
俄罗斯初等数学万题选——三角卷	2012—11	38.00	222
俄罗斯初等数学万题选——代数卷	2013—08	68.00	225
俄罗斯初等数学万题选——几何卷	2014—01	68.00	226
463个俄罗斯几何老问题	2012—01	28.00	152
超越吉米多维奇.数列的极限	2009—11	48.00	58
超越普里瓦洛夫.留数卷	2015—01	28.00	437
超越普里瓦洛夫.无穷乘积与它对解析函数的应用卷	2015—05	28.00	477
超越普里瓦洛夫.积分卷	2015—06	18.00	481
超越普里瓦洛夫.基础知识卷	2015—06	28.00	482
超越普里瓦洛夫.数项级数卷	2015—07	38.00	489
初等数论难题集(第一卷)	2009—05	68.00	44
初等数论难题集(第二卷)(上、下)	2011—02	128.00	82,83
数论概貌	2011—03	18.00	93
代数数论(第二版)	2013—08	58.00	94
代数多项式	2014—06	38.00	289
初等数论的知识与问题	2011—02	28.00	95
超越数论基础	2011—03	28.00	96
数论初等教程	2011—03	28.00	97
数论基础	2011—03	18.00	98
数论基础与维诺格拉多夫	2014—03	18.00	292

Ⅲ

哈尔滨工业大学出版社刘培杰数学工作室
已出版(即将出版)图书目录

书 名	出版时间	定 价	编号
解析数论基础	2012—08	28.00	216
解析数论基础(第二版)	2014—01	48.00	287
解析数论问题集(第二版)	2014—05	88.00	343
数论入门	2011—03	38.00	99
代数数论入门	2015—03	38.00	448
数论开篇	2012—07	28.00	194
解析数论引论	2011—03	48.00	100
Barban Davenport Halberstam 均值和	2009—01	40.00	33
基础数论	2011—03	28.00	101
初等数论100例	2011—05	18.00	122
初等数论经典例题	2012—07	18.00	204
最新世界各国数学奥林匹克中的初等数论试题(上、下)	2012—01	138.00	144,145
初等数论(Ⅰ)	2012—01	18.00	156
初等数论(Ⅱ)	2012—01	18.00	157
初等数论(Ⅲ)	2012—01	28.00	158
平面几何与数论中未解决的新老问题	2013—01	68.00	229
代数数论简史	2014—11	28.00	408
代数数论	2015—09	88.00	532
数论导引提要及习题解答	2016—01	48.00	559
谈谈素数	2011—03	18.00	91
平方和	2011—03	18.00	92
复变函数引论	2013—10	68.00	269
伸缩变换与抛物旋转	2015—01	38.00	449
无穷分析引论(上)	2013—04	88.00	247
无穷分析引论(下)	2013—04	98.00	245
数学分析	2014—04	28.00	338
数学分析中的一个新方法及其应用	2013—01	38.00	231
数学分析例选:通过范例学技巧	2013—01	88.00	243
高等代数例选:通过范例学技巧	2015—06	88.00	475
三角级数论(上册)(陈建功)	2013—01	38.00	232
三角级数论(下册)(陈建功)	2013—01	48.00	233
三角级数论(哈代)	2013—06	48.00	254
三角级数	2015—07	28.00	263
超越数	2011—03	18.00	109
三角和方法	2011—03	18.00	112
整数论	2011—05	38.00	120
从整数谈起	2015—10	28.00	538
随机过程(Ⅰ)	2014—01	78.00	224
随机过程(Ⅱ)	2014—01	68.00	235
算术探索	2011—12	158.00	148
组合数学	2012—04	28.00	178
组合数学浅谈	2012—03	28.00	159
丢番图方程引论	2012—03	48.00	172
拉普拉斯变换及其应用	2015—02	38.00	447
高等代数.上	2016—01	38.00	548
高等代数.下	2016—01	38.00	549

哈尔滨工业大学出版社刘培杰数学工作室
已出版(即将出版)图书目录

书　　名	出版时间	定　价	编号
高等代数教程	2016—01	58.00	579
数学解析教程.上卷.1	2016—01	58.00	546
数学解析教程.上卷.2	2016—01	38.00	553
函数构造论.上	2016—01	38.00	554
函数构造论.下	即将出版		555
数与多项式	2016—01	38.00	558
概周期函数	2016—01	48.00	572
变叙的项的极限分布律	2016—01	18.00	573
整函数	2012—08	18.00	161
近代拓扑学研究	2013—04	38.00	239
多项式和无理数	2008—01	68.00	22
模糊数据统计学	2008—03	48.00	31
模糊分析学与特殊泛函空间	2013—01	68.00	241
谈谈不定方程	2011—05	28.00	119
常微分方程	2016—01	58.00	586
平稳随机函数导论	2016—03	48.00	587
量子力学原理·上	2016—01	38.00	588
受控理论与解析不等式	2012—05	78.00	165
解析不等式新论	2009—06	68.00	48
建立不等式的方法	2011—03	98.00	104
数学奥林匹克不等式研究	2009—08	68.00	56
不等式研究(第二辑)	2012—02	68.00	153
不等式的秘密(第一卷)	2012—02	28.00	154
不等式的秘密(第一卷)(第2版)	2014—02	38.00	286
不等式的秘密(第二卷)	2014—01	38.00	268
初等不等式的证明方法	2010—06	38.00	123
初等不等式的证明方法(第二版)	2014—11	38.00	407
不等式·理论·方法(基础卷)	2015—07	38.00	496
不等式·理论·方法(经典不等式卷)	2015—07	38.00	497
不等式·理论·方法(特殊类型不等式卷)	2015—07	48.00	498
同余理论	2012—05	38.00	163
[x]与{x}	2015—04	48.00	476
极值与最值.上卷	2015—06	28.00	486
极值与最值.中卷	2015—06	38.00	487
极值与最值.下卷	2015—06	28.00	488
整数的性质	2012—11	38.00	192
多项式理论	2015—10	88.00	541
历届美国中学生数学竞赛试题及解答(第一卷)1950—1954	2014—07	18.00	277
历届美国中学生数学竞赛试题及解答(第二卷)1955—1959	2014—04	18.00	278
历届美国中学生数学竞赛试题及解答(第三卷)1960—1964	2014—06	18.00	279
历届美国中学生数学竞赛试题及解答(第四卷)1965—1969	2014—04	28.00	280
历届美国中学生数学竞赛试题及解答(第五卷)1970—1972	2014—06	18.00	281
历届美国中学生数学竞赛试题及解答(第七卷)1981—1986	2015—01	18.00	424

哈尔滨工业大学出版社刘培杰数学工作室
已出版(即将出版)图书目录

书 名	出版时间	定 价	编号
历届 IMO 试题集(1959—2005)	2006—05	58.00	5
历届 CMO 试题集	2008—09	28.00	40
历届中国数学奥林匹克试题集	2014—10	38.00	394
历届加拿大数学奥林匹克试题集	2012—08	38.00	215
历届美国数学奥林匹克试题集:多解推广加强	2012—08	38.00	209
历届波兰数学竞赛试题集.第1卷,1949~1963	2015—03	18.00	453
历届波兰数学竞赛试题集.第2卷,1964~1976	2015—03	18.00	454
历届巴尔干数学奥林匹克试题集	2015—05	38.00	466
保加利亚数学奥林匹克	2014—10	38.00	393
圣彼得堡数学奥林匹克试题集	2015—01	38.00	429
历届国际大学生数学竞赛试题集(1994—2010)	2012—01	28.00	143
全国大学生数学夏令营数学竞赛试题及解答	2007—03	28.00	15
全国大学生数学竞赛辅导教程	2012—07	28.00	189
全国大学生数学竞赛复习全书	2014—04	48.00	340
历届美国大学生数学竞赛试题集	2009—03	88.00	43
前苏联大学生数学奥林匹克竞赛题解(上编)	2012—04	28.00	169
前苏联大学生数学奥林匹克竞赛题解(下编)	2012—04	38.00	170
历届美国数学邀请赛试题集	2014—01	48.00	270
全国高中数学竞赛试题及解答.第1卷	2014—07	38.00	331
大学生数学竞赛讲义	2014—09	28.00	371
亚太地区数学奥林匹克竞赛题	2015—07	18.00	492
高考数学临门一脚(含密押三套卷)(理科版)	2015—01	24.80	421
高考数学临门一脚(含密押三套卷)(文科版)	2015—01	24.80	422
新课标高考数学题型全归纳(文科版)	2015—05	72.00	467
新课标高考数学题型全归纳(理科版)	2015—05	82.00	468
王连笑教你怎样学数学:高考选择题解题策略与客观题实用训练	2014—01	48.00	262
王连笑教你怎样学数学:高考数学高层次讲座	2015—02	48.00	432
高考数学的理论与实践	2009—08	38.00	53
高考数学核心题型解题方法与技巧	2010—01	28.00	86
高考思维新平台	2014—03	38.00	259
30分钟拿下高考数学选择题、填空题(第二版)	2012—01	28.00	146
高考数学压轴题解题诀窍(上)	2012—02	78.00	166
高考数学压轴题解题诀窍(下)	2012—03	28.00	167
北京市五区文科数学三年高考模拟题详解:2013~2015	2015—08	48.00	500
北京市五区理科数学三年高考模拟题详解:2013~2015	2015—09	68.00	505
向量法巧解数学高考题	2009—08	28.00	54
高考数学万能解题法	2015—09	28.00	534
高考物理万能解题法	2015—09	28.00	537
高考化学万能解题法	2015—11	25.00	557
我一定要赚分:高中物理	2016—01	38.00	580
数学高考参考	2016—01	78.00	589
2011~2015年全国及各省市高考数学文科精品试题审题要津与解法研究	2015—10	68.00	539
2011~2015年全国及各省市高考数学理科精品试题审题要津与解法研究	2015—10	88.00	540
最新全国及各省市高考数学试卷解法研究及点拨评析	2009—02	38.00	41
2011年全国及各省市高考数学试题审题要津与解法研究	2011—10	48.00	139

哈尔滨工业大学出版社刘培杰数学工作室
已出版(即将出版)图书目录

书　名	出版时间	定　价	编号
2013年全国及各省市高考数学试题解析与点评	2014—01	48.00	282
全国及各省市高考数学试题审题要津与解法研究	2015—02	48.00	450
新课标高考数学——五年试题分章详解(2007～2011)(上、下)	2011—10	78.00	140,141
全国中考数学压轴题审题要津与解法研究	2013—04	78.00	248
新编全国及各省市中考数学压轴题审题要津与解法研究	2014—05	58.00	342
全国及各省市5年中考数学压轴题审题要津与解法研究	2015—04	58.00	462
中考数学专题总复习	2007—04	28.00	6
中考数学较难题、难题常考题型解题方法与技巧.上	2016—01	48.00	584
中考数学较难题、难题常考题型解题方法与技巧.下	2016—01	58.00	585
数学奥林匹克在中国	2014—06	98.00	344
数学奥林匹克问题集	2014—01	38.00	267
数学奥林匹克不等式散论	2010—06	38.00	124
数学奥林匹克不等式欣赏	2011—09	38.00	138
数学奥林匹克超级题库(初中卷上)	2010—01	58.00	66
数学奥林匹克不等式证明方法和技巧(上、下)	2011—08	158.00	134,135
新编640个世界著名数学智力趣题	2014—01	88.00	242
500个最新世界著名数学智力趣题	2008—06	48.00	3
400个最新世界著名数学最值问题	2008—09	48.00	36
500个世界著名数学征解问题	2009—06	48.00	52
400个中国最佳初等数学征解老问题	2010—01	48.00	60
500个俄罗斯数学经典老题	2011—01	28.00	81
1000个国外中学物理好题	2012—04	48.00	174
300个日本高考数学题	2012—05	38.00	142
500个前苏联早期高考数学试题及解答	2012—05	28.00	185
546个早期俄罗斯大学生数学竞赛题	2014—03	38.00	285
548个来自美苏的数学好问题	2014—11	28.00	396
20所苏联著名大学早期入学试题	2015—02	18.00	452
161道德国工科大学生必做的微分方程习题	2015—05	28.00	469
500个德国工科大学生必做的高数习题	2015—06	28.00	478
德国讲义日本考题.微积分卷	2015—04	48.00	456
德国讲义日本考题.微分方程卷	2015—04	38.00	457
几何变换(Ⅰ)	2014—07	28.00	353
几何变换(Ⅱ)	2015—06	28.00	354
几何变换(Ⅲ)	2015—01	38.00	355
几何变换(Ⅳ)	2015—12	38.00	356
中国初等数学研究　2009卷(第1辑)	2009—05	20.00	45
中国初等数学研究　2010卷(第2辑)	2010—05	30.00	68
中国初等数学研究　2011卷(第3辑)	2011—07	60.00	127
中国初等数学研究　2012卷(第4辑)	2012—07	48.00	190
中国初等数学研究　2014卷(第5辑)	2014—02	48.00	288
中国初等数学研究　2015卷(第6辑)	2015—06	68.00	493
博弈论精粹	2008—03	58.00	30
博弈论精粹.第二版(精装)	2015—01	88.00	461
数学 我爱你	2008—01	28.00	20

哈尔滨工业大学出版社刘培杰数学工作室
已出版(即将出版)图书目录

书　名	出版时间	定　价	编号
精神的圣徒　别样的人生——60位中国数学家成长的历程	2008—09	48.00	39
数学史概论	2009—06	78.00	50
数学史概论(精装)	2013—03	158.00	272
数学史选讲	2016—01	48.00	544
斐波那契数列	2010—02	28.00	65
数学拼盘和斐波那契魔方	2010—07	38.00	72
斐波那契数列欣赏	2011—01	28.00	160
数学的创造	2011—02	48.00	85
数海拾贝	2016—01	48.00	590
数学中的美	2011—02	38.00	84
数论中的美学	2014—12	38.00	351
数学王者　科学巨人——高斯	2015—01	28.00	428
振兴祖国数学的圆梦之旅:中国初等数学研究史话	2015—06	78.00	490
二十世纪中国数学史料研究	2015—10	48.00	536
数字谜、数阵图与棋盘覆盖	2016—01	58.00	298
时间的形状	2016—01	38.00	556
数学解题——靠数学思想给力(上)	2011—07	38.00	131
数学解题——靠数学思想给力(中)	2011—07	48.00	132
数学解题——靠数学思想给力(下)	2011—07	38.00	133
我怎样解题	2013—01	48.00	227
数学解题中的物理方法	2011—06	28.00	114
数学解题的特殊方法	2011—06	48.00	115
中学数学计算技巧	2012—01	48.00	116
中学数学证明方法	2012—01	58.00	117
数学趣题巧解	2012—03	28.00	128
高中数学教学通鉴	2015—05	58.00	479
和高中生漫谈:数学与哲学的故事	2014—08	28.00	369
自主招生考试中的参数方程问题	2015—01	28.00	435
自主招生考试中的极坐标问题	2015—04	28.00	463
近年全国重点大学自主招生数学试题全解及研究.华约卷	2015—02	38.00	441
近年全国重点大学自主招生数学试题全解及研究.北约卷	即将出版		
自主招生数学解证宝典	2015—09	48.00	535
格点和面积	2012—07	18.00	191
射影几何趣谈	2012—04	28.00	175
斯潘纳尔引理——从一道加拿大数学奥林匹克试题谈起	2014—01	28.00	228
李普希兹条件——从几道近年高考数学试题谈起	2012—10	18.00	221
拉格朗日中值定理——从一道北京高考试题的解法谈起	2015—10	18.00	197
闵科夫斯基定理——从一道清华大学自主招生试题谈起	2014—01	28.00	198
哈尔测度——从一道冬令营试题的背景谈起	2012—08	28.00	202
切比雪夫逼近问题——从一道中国台北数学奥林匹克试题谈起	2013—04	38.00	238
伯恩斯坦多项式与贝齐尔曲面——从一道全国高中数学联赛试题谈起	2013—03	38.00	236
卡塔兰猜想——从一道普特南竞赛试题谈起	2013—06	18.00	256

哈尔滨工业大学出版社刘培杰数学工作室
已出版(即将出版)图书目录

书　名	出版时间	定　价	编号
麦卡锡函数和阿克曼函数——从一道前南斯拉夫数学奥林匹克试题谈起	2012—08	18.00	201
贝蒂定理与拉姆贝克莫尔定理——从一个拣石子游戏谈起	2012—08	18.00	217
皮亚诺曲线和豪斯道夫分球定理——从无限集谈起	2012—08	18.00	211
平面凸图形与凸多面体	2012—10	28.00	218
斯坦因豪斯问题——从一道二十五省市自治区中学数学竞赛试题谈起	2012—07	18.00	196
纽结理论中的亚历山大多项式与琼斯多项式——从一道北京市高一数学竞赛试题谈起	2012—07	28.00	195
原则与策略——从波利亚"解题表"谈起	2013—04	38.00	244
转化与化归——从三大尺规作图不能问题谈起	2012—08	28.00	214
代数几何中的贝祖定理(第一版)——从一道IMO试题的解法谈起	2013—08	18.00	193
成功连贯理论与约当块理论——从一道比利时数学竞赛试题谈起	2012—04	18.00	180
磨光变换与范·德·瓦尔登猜想——从一道环球城市竞赛试题谈起	即将出版		
素数判定与大数分解	2014—08	18.00	199
置换多项式及其应用	2012—10	18.00	220
椭圆函数与模函数——从一道美国加州大学洛杉矶分校(UCLA)博士资格考题谈起	2012—10	28.00	219
差分方程的拉格朗日方法——从一道2011年全国高考理科试题的解法谈起	2012—08	28.00	200
力学在几何中的一些应用	2013—01	38.00	240
高斯散度定理、斯托克斯定理和平面格林定理——从一道国际大学生数学竞赛试题谈起	即将出版		
康托洛维奇不等式——从一道全国高中联赛试题谈起	2013—03	28.00	337
西格尔引理——从一道第18届IMO试题的解法谈起	即将出版		
罗斯定理——从一道前苏联数学竞赛试题谈起	即将出版		
拉克斯定理和阿廷定理——从一道IMO试题的解法谈起	2014—01	58.00	246
毕卡大定理——从一道美国大学数学竞赛试题谈起	2014—07	18.00	350
贝齐尔曲线——从一道全国高中联赛试题谈起	即将出版		
拉格朗日乘子定理——从一道2005年全国高中联赛试题的高等数学解法谈起	2015—05	28.00	480
雅可比定理——从一道日本数学奥林匹克试题谈起	2013—04	48.00	249
李天岩—约克定理——从一道波兰数学竞赛试题谈起	2014—06	28.00	349
整系数多项式因式分解的一般方法——从克朗耐克算法谈起	即将出版		
布劳维不动点定理——从一道前苏联数学奥林匹克试题谈起	2014—01	38.00	273
压缩不动点定理——从一道高考数学试题的解法谈起	即将出版		
伯恩赛德定理——从一道英国数学奥林匹克试题谈起	即将出版		
布查特—莫斯特定理——从一道上海市初中竞赛试题谈起	即将出版		
数论中的同余数问题——从一道普特南竞赛试题谈起	即将出版		
范·德蒙行列式——从一道美国数学奥林匹克试题谈起	即将出版		
中国剩余定理:总数法构建中国历史年表	2015—01	28.00	430
牛顿程序与方程求根——从一道全国高考试题解法谈起	即将出版		
库默尔定理——从一道IMO预选试题谈起	即将出版		

哈尔滨工业大学出版社刘培杰数学工作室
已出版(即将出版)图书目录

书　名	出版时间	定　价	编号
卢丁定理——从一道冬令营试题的解法谈起	即将出版		
沃斯滕霍姆定理——从一道IMO预选试题谈起	即将出版		
卡尔松不等式——从一道莫斯科数学奥林匹克试题谈起	即将出版		
信息论中的香农熵——从一道近年高考压轴题谈起	即将出版		
约当不等式——从一道希望杯竞赛试题谈起	即将出版		
拉比诺维奇定理	即将出版		
刘维尔定理——从一道《美国数学月刊》征解问题的解法谈起	即将出版		
卡塔兰恒等式与级数求和——从一道IMO试题的解法谈起	即将出版		
勒让德猜想与素数分布——从一道爱尔兰竞赛试题谈起	即将出版		
天平称重与信息论——从一道基辅市数学奥林匹克试题谈起	即将出版		
哈密尔顿-凯莱定理:从一道高中数学联赛试题的解法谈起	2014-09	18.00	376
艾思特曼定理——从一道CMO试题的解法谈起	即将出版		
一个爱尔特希问题——从一道西德数学奥林匹克试题谈起	即将出版		
有限群中的爱丁格尔问题——从一道北京市初中二年级数学竞赛试题谈起	即将出版		
贝克码与编码理论——从一道全国高中联赛试题谈起	即将出版		
帕斯卡三角形	2014-03	18.00	294
蒲丰投针问题——从2009年清华大学的一道自主招生试题谈起	2014-01	38.00	295
斯图姆定理——从一道"华约"自主招生试题的解法谈起	2014-01	18.00	296
许瓦兹引理——从一道加利福尼亚大学伯克利分校数学系博士生试题谈起	2014-08	18.00	297
拉姆塞定理——从王诗宬院士的一个问题谈起	2014-01		299
坐标法	2013-12	28.00	332
数论三角形	2014-04	38.00	341
毕克定理	2014-07	18.00	352
数林掠影	2014-09	48.00	389
我们周围的概率	2014-10	38.00	390
凸函数最值定理:从一道华约自主招生题的解法谈起	2014-10	28.00	391
易学与数学奥林匹克	2014-10	38.00	392
生物数学趣谈	2015-01	18.00	409
反演	2015-01		420
因式分解与圆锥曲线	2015-01	18.00	426
轨迹	2015-01	28.00	427
面积原理:从常庚哲命的一道CMO试题的积分解法谈起	2015-01	48.00	431
形形色色的不动点定理:从一道28届IMO试题谈起	2015-01	38.00	439
柯西函数方程:从一道上海交大自主招生的试题谈起	2015-02	28.00	440
三角恒等式	2015-02	28.00	442
无理性判定:从一道2014年"北约"自主招生试题谈起	2015-01	38.00	443
数学归纳法	2015-03	18.00	451
极端原理与解题	2015-04	28.00	464
法雷级数	2014-08	18.00	367
摆线族	2015-01	38.00	438
函数方程及其解法	2015-05	38.00	470
含参数的方程和不等式	2012-09	28.00	213
希尔伯特第十问题	2016-01	38.00	543
无穷小量的求和	2016-01	28.00	545
切比雪夫多项式:从一道清华大学金秋营试题谈起	2016-01	38.00	583

哈尔滨工业大学出版社刘培杰数学工作室
已出版(即将出版)图书目录

书　　名	出版时间	定　价	编号
中等数学英语阅读文选	2006—12	38.00	13
统计学专业英语	2007—03	28.00	16
统计学专业英语(第二版)	2012—07	48.00	176
统计学专业英语(第三版)	2015—04	68.00	465
幻方和魔方(第一卷)	2012—05	68.00	173
尘封的经典——初等数学经典文献选读(第一卷)	2012—07	48.00	205
尘封的经典——初等数学经典文献选读(第二卷)	2012—07	38.00	206
代换分析:英文	2015—07	38.00	499
实变函数论	2012—06	78.00	181
复变函数论	2015—08	38.00	504
非光滑优化及其变分分析	2014—01	48.00	230
疏散的马尔科夫链	2014—01	58.00	266
马尔科夫过程论基础	2015—01	28.00	433
初等微分拓扑学	2012—07	18.00	182
方程式论	2011—03	38.00	105
初级方程式论	2011—03	28.00	106
Galois 理论	2011—03	18.00	107
古典数学难题与伽罗瓦理论	2012—11	58.00	223
伽罗华与群论	2014—01	28.00	290
代数方程的根式解及伽罗瓦理论	2011—03	28.00	108
代数方程的根式解及伽罗瓦理论(第二版)	2015—01	28.00	423
线性偏微分方程讲义	2011—03	18.00	110
几类微分方程数值方法的研究	2015—05	38.00	485
N 体问题的周期解	2011—03	28.00	111
代数方程式论	2011—05	18.00	121
动力系统的不变量与函数方程	2011—07	48.00	137
基于短语评价的翻译知识获取	2012—02	48.00	168
应用随机过程	2012—04	48.00	187
概率论导引	2012—04	18.00	179
矩阵论(上)	2013—06	58.00	250
矩阵论(下)	2013—06	48.00	251
对称锥互补问题的内点法:理论分析与算法实现	2014—08	68.00	368
抽象代数:方法导引	2013—06	38.00	257
集论	2016—01	48.00	576
多项式理论研究综述	2016—01	38.00	577
函数论	2014—11	78.00	395
反问题的计算方法及应用	2011—11	28.00	147
初等数学研究(Ⅰ)	2008—09	68.00	37
初等数学研究(Ⅱ)(上、下)	2009—05	118.00	46,47
数阵及其应用	2012—02	28.00	164
绝对值方程—折边与组合图形的解析研究	2012—07	48.00	186
代数函数论(上)	2015—07	38.00	494
代数函数论(下)	2015—07	38.00	495
偏微分方程论:法文	2015—10	48.00	533
闵嗣鹤文集	2011—03	98.00	102
吴从炘数学活动三十年(1951~1980)	2010—07	99.00	32
吴从炘数学活动又三十年(1981~2010)	2015—07	98.00	491

哈尔滨工业大学出版社刘培杰数学工作室
已出版(即将出版)图书目录

书　　名	出版时间	定　价	编号
趣味初等方程妙题集锦	2014—09	48.00	388
趣味初等数论选美与欣赏	2015—02	48.00	445
耕读笔记(上卷):一位农民数学爱好者的初数探索	2015—04	28.00	459
耕读笔记(中卷):一位农民数学爱好者的初数探索	2015—05	28.00	483
耕读笔记(下卷):一位农民数学爱好者的初数探索	2015—05	28.00	484
几何不等式研究与欣赏.上卷	2016—01	88.00	547
几何不等式研究与欣赏.下卷	2016—01	48.00	552
初等数列研究与欣赏·上	2016—01	48.00	570
初等数列研究与欣赏·下	2016—01	48.00	571
数贝偶拾——高考数学题研究	2014—04	28.00	274
数贝偶拾——初等数学研究	2014—04	38.00	275
数贝偶拾——奥数题研究	2014—04	48.00	276
集合、函数与方程	2014—01	28.00	300
数列与不等式	2014—01	38.00	301
三角与平面向量	2014—01	28.00	302
平面解析几何	2014—01	38.00	303
立体几何与组合	2014—01	28.00	304
极限与导数、数学归纳法	2014—01	38.00	305
趣味数学	2014—03	28.00	306
教材教法	2014—04	68.00	307
自主招生	2014—05	58.00	308
高考压轴题(上)	2015—01	48.00	309
高考压轴题(下)	2014—10	68.00	310
从费马到怀尔斯——费马大定理的历史	2013—10	198.00	I
从庞加莱到佩雷尔曼——庞加莱猜想的历史	2013—10	298.00	II
从切比雪夫到爱尔特希(上)——素数定理的初等证明	2013—07	48.00	III
从切比雪夫到爱尔特希(下)——素数定理100年	2012—12	98.00	III
从高斯到盖尔方特——二次域的高斯猜想	2013—10	198.00	IV
从库默尔到朗兰兹——朗兰兹猜想的历史	2014—01	98.00	V
从比勃巴赫到德布朗斯——比勃巴赫猜想的历史	2014—02	298.00	VI
从麦比乌斯到陈省身——麦比乌斯变换与麦比乌斯带	2014—02	298.00	VII
从布尔到豪斯道夫——布尔方程与格论漫谈	2013—10	198.00	VIII
从开普勒到阿诺德——三体问题的历史	2014—05	298.00	IX
从华林到华罗庚——华林问题的历史	2013—10	298.00	X
第19~23届"希望杯"全国数学邀请赛试题审题要津详细评注(初一版)	2014—03	28.00	333
第19~23届"希望杯"全国数学邀请赛试题审题要津详细评注(初二、初三版)	2014—03	38.00	334
第19~23届"希望杯"全国数学邀请赛试题审题要津详细评注(高一版)	2014—03	28.00	335
第19~23届"希望杯"全国数学邀请赛试题审题要津详细评注(高二版)	2014—03	38.00	336
第19~25届"希望杯"全国数学邀请赛试题审题要津详细评注(初一版)	2015—01	38.00	416
第19~25届"希望杯"全国数学邀请赛试题审题要津详细评注(初二、初三版)	2015—01	58.00	417
第19~25届"希望杯"全国数学邀请赛试题审题要津详细评注(高一版)	2015—01	48.00	418
第19~25届"希望杯"全国数学邀请赛试题审题要津详细评注(高二版)	2015—01	48.00	419

哈尔滨工业大学出版社刘培杰数学工作室
已出版(即将出版)图书目录

书 名	出版时间	定 价	编号
吴振奎高等数学解题真经(概率统计卷)	2012—01	38.00	149
吴振奎高等数学解题真经(微积分卷)	2012—01	68.00	150
吴振奎高等数学解题真经(线性代数卷)	2012—01	58.00	151
钱昌本教你快乐学数学(上)	2011—12	48.00	155
钱昌本教你快乐学数学(下)	2012—03	58.00	171
高等数学解题全攻略(上卷)	2013—06	58.00	252
高等数学解题全攻略(下卷)	2013—06	58.00	253
高等数学复习纲要	2014—01	18.00	384
三角函数	2014—01	38.00	311
不等式	2014—01	38.00	312
数列	2014—01	38.00	313
方程	2014—01	28.00	314
排列和组合	2014—01	28.00	315
极限与导数	2014—01	28.00	316
向量	2014—09	38.00	317
复数及其应用	2014—08	28.00	318
函数	2014—01	38.00	319
集合	即将出版		320
直线与平面	2014—01	28.00	321
立体几何	2014—04	28.00	322
解三角形	即将出版		323
直线与圆	2014—01	28.00	324
圆锥曲线	2014—01	38.00	325
解题通法(一)	2014—07	38.00	326
解题通法(二)	2014—07	38.00	327
解题通法(三)	2014—05	38.00	328
概率与统计	2014—01	28.00	329
信息迁移与算法	即将出版		330
物理奥林匹克竞赛大题典——力学卷	2014—11	48.00	405
物理奥林匹克竞赛大题典——热学卷	2014—04	28.00	339
物理奥林匹克竞赛大题典——电磁学卷	2015—07	48.00	406
物理奥林匹克竞赛大题典——光学与近代物理卷	2014—06	28.00	345
历届中国东南地区数学奥林匹克试题集(2004~2012)	2014—06	18.00	346
历届中国西部地区数学奥林匹克试题集(2001~2012)	2014—07	18.00	347
历届中国女子数学奥林匹克试题集(2002~2012)	2014—08	18.00	348
美国高中数学竞赛五十讲.第1卷(英文)	2014—08	28.00	357
美国高中数学竞赛五十讲.第2卷(英文)	2014—08	28.00	358
美国高中数学竞赛五十讲.第3卷(英文)	2014—09	28.00	359
美国高中数学竞赛五十讲.第4卷(英文)	2014—09	28.00	360
美国高中数学竞赛五十讲.第5卷(英文)	2014—10	28.00	361
美国高中数学竞赛五十讲.第6卷(英文)	2014—11	28.00	362
美国高中数学竞赛五十讲.第7卷(英文)	2014—12	28.00	363
美国高中数学竞赛五十讲.第8卷(英文)	2015—01	28.00	364
美国高中数学竞赛五十讲.第9卷(英文)	2015—01	28.00	365
美国高中数学竞赛五十讲.第10卷(英文)	2015—02	38.00	366

哈尔滨工业大学出版社刘培杰数学工作室
已出版(即将出版)图书目录

书　　名	出版时间	定　价	编号
IMO 50 年.第 1 卷(1959—1963)	2014—11	28.00	377
IMO 50 年.第 2 卷(1964—1968)	2014—11	28.00	378
IMO 50 年.第 3 卷(1969—1973)	2014—09	28.00	379
IMO 50 年.第 4 卷(1974—1978)	即将出版		380
IMO 50 年.第 5 卷(1979—1984)	2015—04	38.00	381
IMO 50 年.第 6 卷(1985—1989)	2015—04	58.00	382
IMO 50 年.第 7 卷(1990—1994)	2016—01	48.00	383
IMO 50 年.第 8 卷(1995—1999)	即将出版		384
IMO 50 年.第 9 卷(2000—2004)	2015—04	58.00	385
IMO 50 年.第 10 卷(2005—2009)	2016—01	48.00	386
历届美国大学生数学竞赛试题集.第一卷(1938—1949)	2015—01	28.00	397
历届美国大学生数学竞赛试题集.第二卷(1950—1959)	2015—01	28.00	398
历届美国大学生数学竞赛试题集.第三卷(1960—1969)	2015—01	28.00	399
历届美国大学生数学竞赛试题集.第四卷(1970—1979)	2015—01	18.00	400
历届美国大学生数学竞赛试题集.第五卷(1980—1989)	2015—01	28.00	401
历届美国大学生数学竞赛试题集.第六卷(1990—1999)	2015—01	28.00	402
历届美国大学生数学竞赛试题集.第七卷(2000—2009)	2015—08	18.00	403
历届美国大学生数学竞赛试题集.第八卷(2010—2012)	2015—01	18.00	404
新课标高考数学创新题解题诀窍:总论	2014—09	28.00	372
新课标高考数学创新题解题诀窍:必修 1～5 分册	2014—08	38.00	373
新课标高考数学创新题解题诀窍:选修 2—1,2—2,1—1,1—2 分册	2014—09	38.00	374
新课标高考数学创新题解题诀窍:选修 2—3,4—4,4—5 分册	2014—09	18.00	375
全国重点大学自主招生英文数学试题全攻略:词汇卷	2015—07	48.00	410
全国重点大学自主招生英文数学试题全攻略:概念卷	2015—01	28.00	411
全国重点大学自主招生英文数学试题全攻略:文章选读卷(上)	即将出版		412
全国重点大学自主招生英文数学试题全攻略:文章选读卷(下)	即将出版		413
全国重点大学自主招生英文数学试题全攻略:试题卷	2015—07	38.00	414
全国重点大学自主招生英文数学试题全攻略:名著欣赏卷	即将出版		415
数学物理大百科全书.第 1 卷	2016—01	418.00	508
数学物理大百科全书.第 2 卷	2016—01	408.00	509
数学物理大百科全书.第 3 卷	2016—01	396.00	510
数学物理大百科全书.第 4 卷	2016—01	408.00	511
数学物理大百科全书.第 5 卷	2016—01	368.00	512
劳埃德数学趣题大全.题目卷.1:英文	2016—01	18.00	516
劳埃德数学趣题大全.题目卷.2:英文	2016—01	18.00	517
劳埃德数学趣题大全.题目卷.3:英文	2016—01	18.00	518
劳埃德数学趣题大全.题目卷.4:英文	2016—01	18.00	519
劳埃德数学趣题大全.题目卷.5:英文	2016—01	18.00	520
劳埃德数学趣题大全.答案卷:英文	2016—01	18.00	521

哈尔滨工业大学出版社刘培杰数学工作室
已出版(即将出版)图书目录

书　　名	出版时间	定　价	编号
李成章教练奥数笔记.第1卷	2016—01	48.00	522
李成章教练奥数笔记.第2卷	2016—01	48.00	523
李成章教练奥数笔记.第3卷	2016—01	38.00	524
李成章教练奥数笔记.第4卷	2016—01	38.00	525
李成章教练奥数笔记.第5卷	2016—01	38.00	526
李成章教练奥数笔记.第6卷	2016—01	38.00	527
李成章教练奥数笔记.第7卷	2016—01	38.00	528
李成章教练奥数笔记.第8卷	2016—01	48.00	529
李成章教练奥数笔记.第9卷	2016—01	28.00	530

书　　名	出版时间	定　价	编号
zeta 函数,q-zeta 函数,相伴级数与积分	2015—08	88.00	513
微分形式:理论与练习	2015—08	58.00	514
离散与微分包含的逼近和优化	2015—08	58.00	515
艾伦·图灵:他的工作与影响	2016—01	98.00	560
测度理论概率导论,第2版	2016—01	88.00	561
带有潜在故障恢复系统的半马尔柯夫模型控制	2016—01	98.00	562
数学分析原理	2016—01	88.00	563
随机偏微分方程的有效动力学	2016—01	88.00	564
图的谱半径	2016—01	58.00	565
量子机器学习中数据挖掘的量子计算方法	2016—01	98.00	566
量子物理的非常规方法	2016—01	118.00	567
运输过程的统一非局部理论:广义波尔兹曼物理动力学,第2版	2016—01	198.00	568
量子力学与经典力学之间的联系在原子、分子及电动力学系统建模中的应用	2016—01	58.00	569

联系地址:哈尔滨市南岗区复华四道街10号　哈尔滨工业大学出版社刘培杰数学工作室
网　　址:http://lpj.hit.edu.cn/
邮　　编:150006
联系电话:0451—86281378　　　13904613167
E-mail:lpj1378@163.com